# A2 Geography for Edexcel



## Teacher's Book

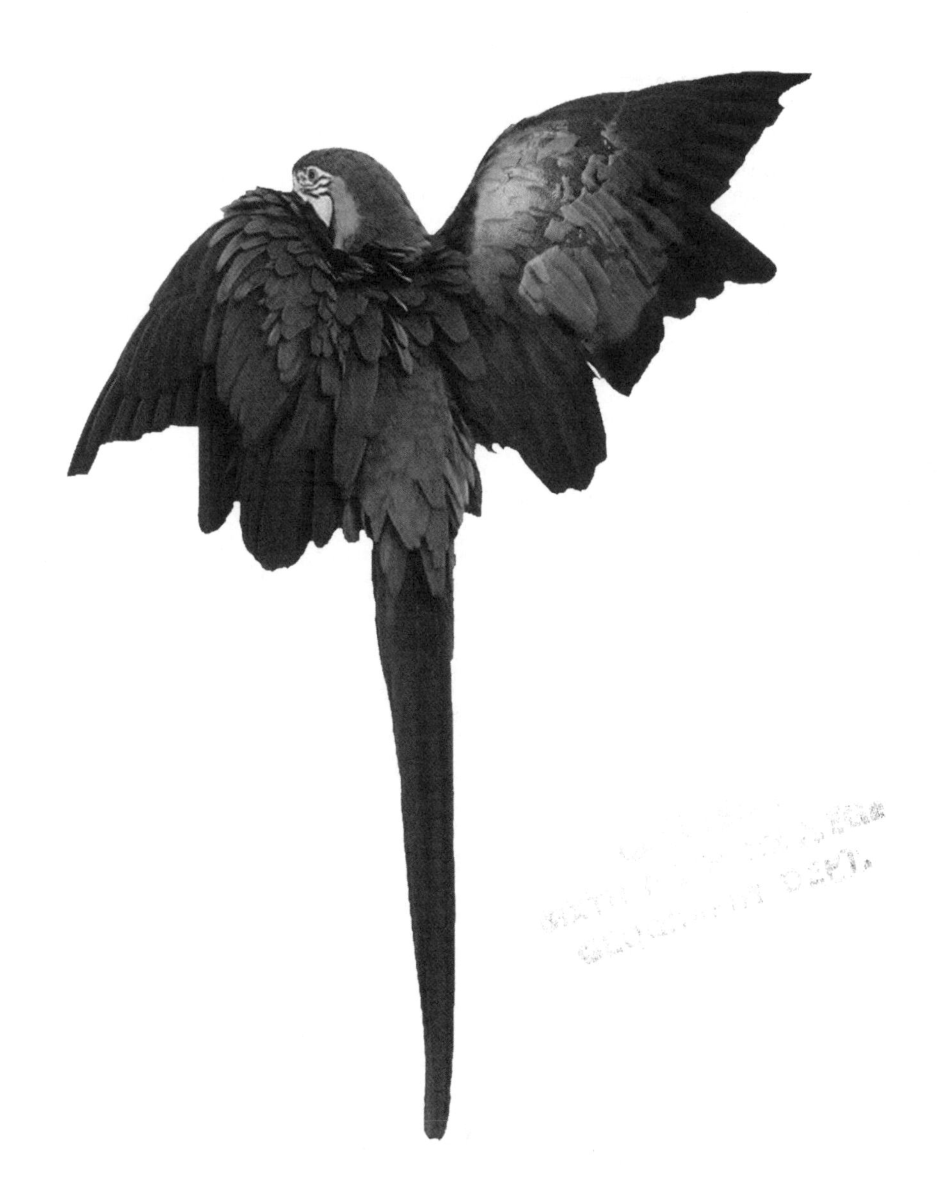

Digby • Hurst

OXFORD

OXFORD
UNIVERSITY PRESS

Great Clarendon Street, Oxford OX2 6DP

Oxford University Press is a department of the University of Oxford.
It furthers the University's objective of excellence in research,
scholarship, and education by publishing worldwide in

Oxford New York

Auckland Cape Town Dar es Salaam Hong Kong Karachi
Kuala Lumpur Madrid Melbourne Mexico City Nairobi
New Delhi Shanghai Taipei Toronto

With offices in

Argentina Austria Brazil Chile Czech Republic France Greece
Guatemala Hungary Italy Japan Poland Portugal Singapore
South Korea Switzerland Thailand Turkey Ukraine Vietnam





Authors: Bob Digby, Catherine Hurst





First published 2009





British Library Cataloguing in Publication Data

Data available

ISBN 978-0-19-913484-7

10 9 8 7 6 5 4 3 2 1

Printed Bell and Bain Ltd., Glasgow


Paper used in the production of this book is a natural, recyclable product made
from wood grown in sustainable forests. The manufacturing process conforms to
the environmental regulations of the country of origin.

# Contents

# About this course

This course has been written to meet the requirements of the Edexcel GCE in Geography. It has been written to make AS and A2 learning accessible, and we hope it will help you and your students succeed.

## The course components

**The Student Books**

- Two books:
  - *AS Geography for Edexcel*, written to meet the requirements of the Edexcel Advanced Subsidiary GCE in Geography.
  - *A2 Geography for Edexcel*, written to meet the requirements of the Edexcel Advanced GCE in Geography.
- All the content and case studies students will need for their course.
- Coverage of all the core and option topics.
- Chapters divided into clearly-identified units of between two and eight pages.
- Aims of the unit given in student-friendly language at the start of each unit.
- 'Over to you' and 'On your own' questions for each unit.
- Exam-style questions for each chapter.
- The first page of each chapter gives the enquiry questions detailed in the specification, and makes it clear to students what they have to know and learn.

**The Teacher's Books**

- One for each Student Book.
- Material to help you for each unit in the Student Books.
- Brief unit overviews.
- Key ideas.
- Unit outcomes.
- Ideas for starters and plenaries.

**The Activities and Planning OxBox CD-ROMs**

- One for the AS part of the course, one for A2.
- Photographs and resources from the Student Books.
- Exam-style questions, with key points for success and mark schemes.
- Support for fieldwork (AS Unit 2) and research (A2 Unit 4).
- Answer guidelines for 'Over to you', 'On your own', and 'What do you think?' questions.
- Editable 'Over to you' and 'On your own' questions.
- Customisable planning materials; the lesson player helps you arrange and launch the resources you want to use in sequence.
- User management facility which allows you to easily import class registers and create user accounts for all your students.

**OxBox technology**: OxBox CD-ROMs are bought individually. The software then allows the content to be tipped together to create a single resource on your network.

They are customisable – you can add your own resources. There are easy-to-follow guidelines on how to do this.

# Using this Teacher's Book

This Teacher's Book is intended to save you time and effort. It offers support for the Student Book, and will help you prepare detailed course and lesson plans.

## What it provides for Core chapters

For each Core chapter in the Student Book (Chapters 1–6), this Teacher's Book provides:

**1** a concise overview of the chapter

**2** support for each unit in the chapter

It also has a glossary at the back, covering the terms students will meet.

## 1 The chapter overview

This is your introduction to the corresponding chapter in the Student Book. This is what it provides:

- **Chapter outline**: the title of each unit, together with a one-line summary of what it covers.
- **About the topic**: a brief summary of the topic, set in its specification context.
- **About the chapter**: a brief summary of the content and approach of the Student Book chapter.
- **Key vocabulary**: a list of key words and terms for the chapter.

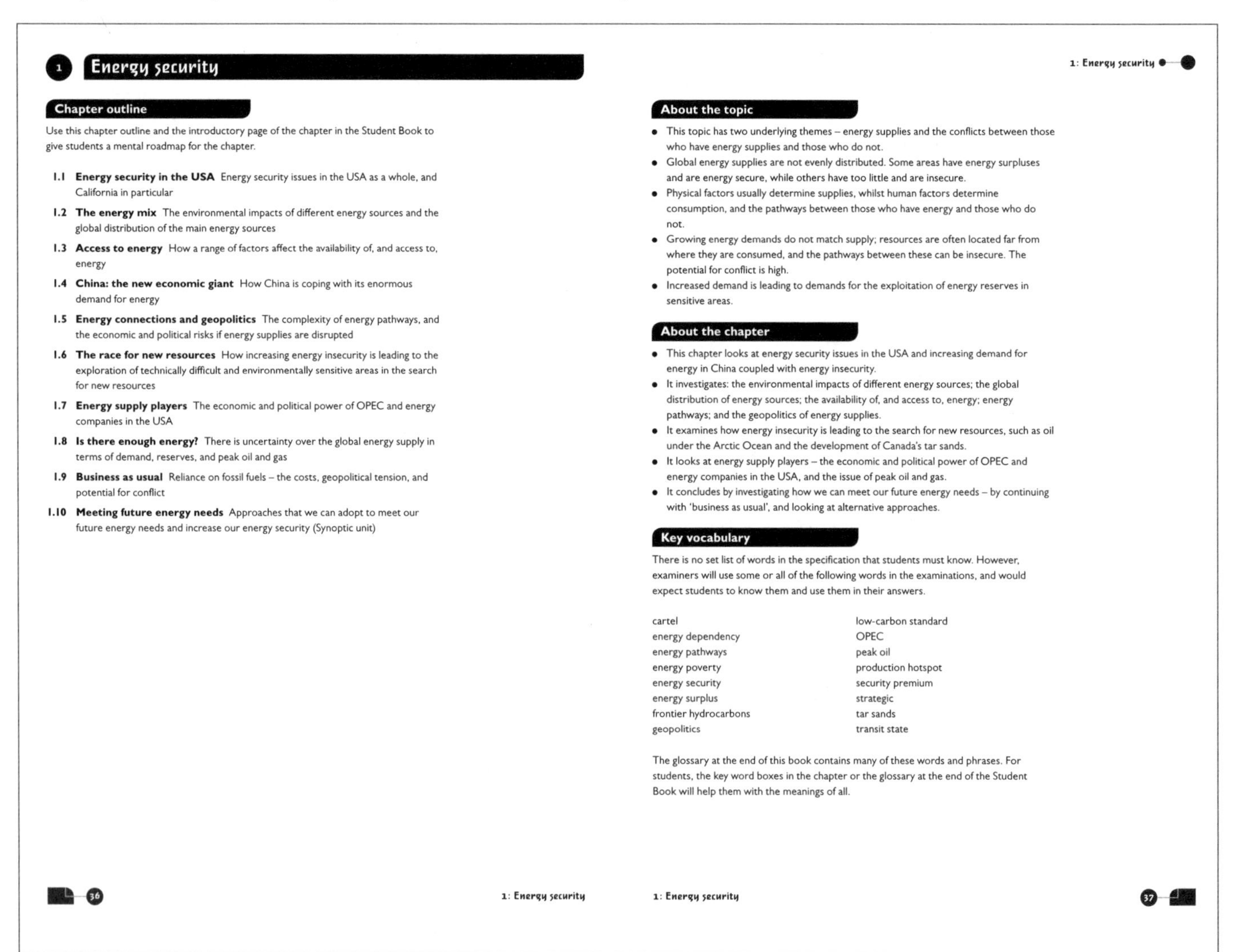

1 **Energy security**

**Chapter outline**

Use this chapter outline and the introductory page of the chapter in the Student Book to give students a mental roadmap for the chapter.

**1.1 Energy security in the USA** Energy security issues in the USA as a whole, and California in particular

**1.2 The energy mix** The environmental impacts of different energy sources and the global distribution of the main energy sources

**1.3 Access to energy** How a range of factors affect the availability of, and access to, energy

**1.4 China: the new economic giant** How China is coping with its enormous demand for energy

**1.5 Energy connections and geopolitics** The complexity of energy pathways, and the economic and political risks if energy supplies are disrupted

**1.6 The race for new resources** How increasing energy insecurity is leading to the exploration of technically difficult and environmentally sensitive areas in the search for new resources

**1.7 Energy supply players** The economic and political power of OPEC and energy companies in the USA

**1.8 Is there enough energy?** There is uncertainty over the global energy supply in terms of demand, reserves, and peak oil and gas

**1.9 Business as usual** Reliance on fossil fuels – the costs, geopolitical tension, and potential for conflict

**1.10 Meeting future energy needs** Approaches that we can adopt to meet our future energy needs and increase our energy security (Synoptic unit)

1: Energy security

**About the topic**

- This topic has two underlying themes – energy supplies and the conflicts between those who have energy supplies and those who do not.
- Global energy supplies are not evenly distributed. Some areas have energy surpluses and are energy secure, while others have too little and are insecure.
- Physical factors usually determine supplies, whilst human factors determine consumption, and the pathways between those who have energy and those who do not.
- Growing energy demands do not match supply; resources are often located far from where they are consumed, and the pathways between these can be insecure. The potential for conflict is high.
- Increased demand is leading to demands for the exploitation of energy reserves in sensitive areas.

**About the chapter**

- This chapter looks at energy security issues in the USA and increasing demand for energy in China coupled with energy insecurity.
- It investigates: the environmental impacts of different energy sources; the global distribution of energy sources; the availability of, and access to, energy; energy pathways; and the geopolitics of energy supplies.
- It examines how energy insecurity is leading to the search for new resources, such as oil under the Arctic Ocean and the development of Canada's tar sands.
- It looks at energy supply players – the economic and political power of OPEC and energy companies in the USA, and the issue of peak oil and gas.
- It concludes by investigating how we can meet our future energy needs – by continuing with 'business as usual', and looking at alternative approaches.

**Key vocabulary**

There is no set list of words in the specification that students must know. However, examiners will use some or all of the following words in the examinations, and would expect students to know them and use them in their answers.

| | |
|---|---|
| cartel | low-carbon standard |
| energy dependency | OPEC |
| energy pathways | peak oil |
| energy poverty | production hotspot |
| energy security | security premium |
| energy surplus | strategic |
| frontier hydrocarbons | tar sands |
| geopolitics | transit state |

The glossary at the end of this book contains many of these words and phrases. For students, the key word boxes in the chapter or the glossary at the end of the Student Book will help them with the meanings of all.

36 1: Energy security

1: Energy security 37

## 2 Support for each unit

These pages give help for each unit in the chapter. This is what they provide:

- **The unit in brief**: tells you what the unit covers, and how it develops.
- **Key ideas**: the key ideas covered in the unit.
- **Unit outcomes**: the expected outcomes for the unit.
- **Ideas for a starter**: several suggestions for a starter.
- **Ideas for plenaries**: several suggestions for plenaries that could be used throughout lessons, not just at the end.

## What it provides for Options chapters

For each Option chapter in the Student Book (Chapters 7–12), this Teacher's Book provides a concise overview of the chapter. For further information on Unit 4 Options, see pages 28–32.

## The chapter overview

This is your introduction to the corresponding chapter in the Student Book. This is what it provides:

- **About the Option:** a brief summary of the topic, set in its specification context.
- **Introducing the Option:** a brief summary of, and introduction to, the Option.
- **Using the chapter:** a brief summary of the content and key points behind the material included in the chapter.

### 7 Tectonic activity and hazards

Chapter 7 in the Student Book is a sample study, which along with Chapters 8–12 is designed to help students select which Option to study. Students can, and should, use these chapters in their research, but they do not form complete courses. For extra resources on this Option, refer to *A2 Geography for Edexcel Activities & Planning OxBox CD-ROM*.

#### About the Option

- Tectonic activity generates a wide range of natural hazards, caused by plate tectonics.
- Tectonics is a key landscape-forming process, producing distinctive landforms in active regions, ranging from minor features such as scarps to vast rift valleys and shield volcanoes.
- Tectonic hazards cause risk to people and their possessions, depending upon their vulnerability, and the magnitude and frequency of the event.
- Risk varies due to factors including level of development, preparedness, and education.
- Hazard impacts may be short-term or long-term.
- People respond to hazard risk in different ways, depending upon their knowledge, technology, and financial resources.

#### Introducing the Option

The movement of the Earth's tectonic plates can be hazardous for human activity. Volcanic eruptions, earthquakes, and tsunami often grab the news headlines when many lives are lost. The short-, medium-, and long-term impacts of these tectonic events vary in relation to the intensity and frequency of the event, and the nature of the location affected. Levels of economic development, methods of prediction and preparation, and population densities often determine the severity of hazardous events. How people cope and recover depends on the scale and nature of the event.

- Some events are dramatic. The 2004 Boxing Day tsunami, which was caused by an earthquake with a magnitude of 9 to 9.3 under the Indian Ocean (near the coast of Sumatra) killed over 150 000 people. Although the earthquake was localised, the effects of the surge of water were felt over a vast area and reconstruction will take years.
- Some events are small scale and go unnoticed by the rest of the world. Minor earth tremors and volcanic eruptions occur throughout the year in Iceland. They usually only cause short-term disturbance and help to generate valuable tourist revenues as people deliberately visit the country to watch the geysers and bathe in geothermally warmed pool, while Icelanders enjoy cheap central heating and year-round salad crops and fruit grown in glass houses heated by the geothermal activity.
- Tectonic activity can also create distinctive landscapes, like the Great African Rift Valley, fault planes, and mountains – as well as trigger other geomorphic hazards, such as landslides, floods, and climatic change.

7: Tectonic activity and hazards

7: Tectonic activity and hazards

#### Using Chapter 7

Chapter 7 contains six pages of resources.

- Page 264 introduces the topic with a case study of Montserrat.
  - Following two years of small eruptions between 1995 and 1997, the Chances Peak volcano fully erupted in June 1997.
  - Although Plymouth (Montserrat's capital) was destroyed, advance preparation and warnings saved many lives.
  - The government's four-year Sustainable Development Plan, launched in 2003, aimed to restore confidence, rebuild infrastructure, and the economy.
- Pages 265–6 investigate the issue further using resources. Students are guided, but will need to think about each resource carefully. The resources look at:
  - Montserrat's revival
  - funding Montserrat's recovery
  - how the risks of living on Montserrat are being managed.
- Page 267 provides background information about tectonic hazards and risks.
  - Disaster was avoided in Montserrat as two-thirds of the island was declared an exclusion zone before the 1997 eruption.
  - The people of Montserrat still face an uncertain future and the island's economy has not recovered.
  - People living in naturally volatile areas are exposed to increasing risks and are considered vulnerable. When vulnerability and hazards coincide, there's a high risk of disaster.
  - As global population grows more people are living in danger zones.
- Pages 268–9 investigate the issue in other areas of the world, again using resources. These look at the impacts of earthquakes in areas with differing population densities – the Chinese earthquake in May 2008, and the effects of earthquakes in Japan.
- Page 270 provides activities, useful websites for further research, films, books, and music on the theme, and examples of the kinds of questions that students will meet in the exam.

Building up a file of tectonic hazards will help students to understand issues about their impacts and management. Useful tectonic events include:

- in 1963, Icelanders were able to witness the birth of Surtsey, a new volcanic island across the mid-Atlantic constructive plate boundary;
- the Pakistan earthquake in Balakot which initially killed 19 000 in October 2005, with a further 60 000 dying in the aftermath following a harsh winter in an area of poverty;
- the largest earthquake in the UK for 25 years on 27 February 2008 (the epicentre of the 5.2 magnitude earthquake was near Market Rasen in Lincolnshire);
- 15 000 Colombians were forced to evacuate as the Nevado del Huila Volcano began to erupt in April 2008; between May and June 2008, the villagers around the Chaitén Volcano in Chile faced outpourings of ash.

# About the Student Book

## Matching the Edexcel specification

*A2 Geography for Edexcel* has been written to meet the requirements of the specification. The chapter and unit headings in the Student Book correspond to the topics in the specification, and we have pursued many of the suggestions in the 'Teaching and Learning' sections in the specification.

## References to the specification

The Enquiry questions and 'What you need to learn' from the specification are detailed on the opening page for each chapter.

The chapter called 'Exams: how to be successful' at the end of the Student Book tells students about the specification and the exams. It tells them how to make the step up to A2 by developing extended writing. It tells them how to prepare for the synoptic resources in Unit 3 and how to research for Unit 4. It also tells them how questions are marked, and how marks are gained and lost.

## Promoting best practice

Geography is very much about using a variety of resources to identify, describe, and explain the often multi-causal nature of geographical phenomena, and the Student Book deliberately offers information and data in a number of ways. We hope the questions there will encourage students to approach diagrams, graphs, data tables, and photographs, and ideas and scenarios, in a questioning manner.

## At the right level

We have tried to ensure that the definitions, descriptions, and explanations are clear and concise, and appropriately pitched for the range of students now embarking on A2 courses.

The material is intended to be accessible to all students, but we have also aimed to provide plenty of opportunity to stretch and challenge stronger students.

The questions cater for the full ability range, and provide plenty of scope for independent learning.

## Interesting and relevant

Information is presented in a lively, thought-provoking way. The case studies are as current as possible at the time of writing, and are supported by good quality photographs. We hope we've achieved the right balance between breadth and depth.

# About the questions in the Student Book

## Questions in Chapters 1–6

There are questions for each unit in the Student Book.

- 'Over to you' questions mostly provide students with opportunities for collaboration, for pair or group work.
- 'On your own' questions mostly provide students with opportunities for independent work.
- Resources and questions are quite closely linked to encourage student learning to be active and enquiring. The questions cater for the full ability range.
- The Student Book also contains 'exam-style' questions, with marks allocated. These are clearly identified. They are included in certain 'On your own' questions, and on the final page for each chapter. They give the chance of valuable exam practice.
- One unit in each chapter is a synoptic unit, intended to help students get to grips with the synoptic requirement of the specification. These units conclude with a synoptic-style question. They are designed to give students a flavour of what examiners could ask.
- There are also 'What do you think?' boxes. These ask questions about controversial issues, and will challenge students' critical thinking. You could use them in a variety of ways – to spark classroom debate and discussion, or for homework. They are often recommended in the 'Ideas for plenaries' given in this book. Questions of this type will not appear in exam papers.
- The wide range of activities should encourage students to approach diagrams, graphs, data tables, and photographs, and ideas and scenarios, in a questioning manner.

## Questions in Chapters 7–12

There are questions for each chapter in the Student Book.

- 'On your own' questions provide students with opportunities for independent work.
- 'Exam-style' questions on the final page of each chapter give valuable exam practice.
- 'What do you think?' boxes. As for Chapters 1–6, these ask questions about controversial issues, and will challenge students' critical thinking. You could use them in a variety of ways – to spark classroom debate and discussion, or for homework. Questions of this type will not appear in exam papers.

## Answers

The *A2 Geography for Edexcel Activities & Planning OxBox CD-ROM* provides answer guidelines for the 'Over to you', 'On your own', and 'What do you think?' questions.

It also provides key points for success and mark schemes for the exam-style questions.

*are the greatest climate crime. Not only will their development produce 100 million tonnes of greenhouse gases a year by 2012, but it will kill off 147 000 square miles of forest which represents the greatest carbon sink in the world.'*

*Should North America rely on Canada's tar sands to provide an alternative source of oil, or look for alternatives to fossil fuels?*

**What do you think?**

**Over to you**

1 a Identify all the players involved in mining tar sands in Canada.
  b Draw up and complete a conflict matrix to identify which players would agree or disagree with each other.
  c Describe what the matrix shows about exploiting the tar sands.

2 Complete a table to assess the costs and benefits of drilling for oil in the Arctic (economic, social, and environmental).

**On your own**

3 Research another example where people are searching for oil in either extreme or environmentally sensitive areas, e.g. prospecting for oil off the north-west coast of Australia, or developing Colorado's oil shales. Use the following framework for your enquiry:
  a How is the oil formed?
  b What issues are involved in extracting the oil?
  c Who is involved and how?
  d What are the conflicts?
  e How far are the conflicts being resolved?

# Matching the specification

## Introducing the course

The A2 Edexcel Geography specification is intentionally broader than the AS specification and framed around concepts which are global in nature. As a rule, it is much less place-specific than AS; there are, for instance, no compulsory place studies such as the Philippines or California. However, place studies always enhance the ways in which students learn concepts, and the Student Book is rich in place studies. These serve to illustrate global themes ensuring that broad global coverage is achieved.

## Unit 3 Contested planet

Unit 3 'Contested Planet' consists of six compulsory topics:

- Energy security
- Water conflicts
- Biodiversity under threat
- Superpower geographies
- Bridging the development gap
- The technological fix?

## Chapter 1 Energy security

This topic focuses on energy security in terms of energy supply and demand, the issues that these create, and the conflicts between those who have energy and those who do not.

Examples with which students should be familiar include energy-rich regions such as the Middle East and Russia, energy insecure regions such as North America, and issues for developing economies as they increase energy consumption.

Students need to know and understand the extent to which the world is energy secure, the impacts of an increasingly energy insecure world, and what future energy scenarios for the world could be.

Examples used in the Student Book to study each key question are as follows:

### 1 Energy supply, demand, and security: To what extent is the world 'energy secure' at present?

- Different energy types, energy classification, and costs
- Global energy reserves: the imbalance in supply and demand, e.g. Europe vs Russian and Middle Eastern supplies; USA
- Energy sources: their distribution globally and in California/USA; energy poverty in Kenya
- Availability of energy in the UK
- China's increasing demand for energy
- Energy insecurity – UK, China, Japan; securing supplies (e.g. in the Middle East)

### 2 The impacts of energy insecurity: What are the potential impacts of an increasingly 'energy insecure' world?

- Connections: the East Siberia–Pacific Ocean oil pipeline; the rise of the Russian energy oligarchies; Iraqi oil and conflict in the Middle East
- Threats to European supplies from Russia
- The search for new resources, e.g. the Arctic and geopolitics (Russia's scramble for control); the environmental costs of these searches (Canadian tar sands)
- TNCs and energy (Enron), and their influence in US politics
- International energy controls, e.g. OPEC

### 3 Energy security and the future: What might the world's energy future be?

- Future energy growth – 'peak oil and gas'
- Projections for energy production and consumption
- Future scenarios:
  - business as usual (reliance on fossil fuels)
  - alternative sources (nuclear, renewables, conservation, recycling)
  - conserving resources and using less
  - global insecurity, e.g. terrorism
  - OPEC and the future
  - the Middle East – Iraq, Iran and oil security

## Unit 3 Contested planet continued

### Chapter 2 Water conflicts

This topic focuses on water conflicts in terms of water supply and demand, the issues that these create, and the conflicts between those who have water and those who do not.

Examples with which students should be familiar include water-surplus regions both within and between countries (e.g. within the USA or Australia, or between Israel and its neighbours), water insecure regions such as the Middle East, and issues for developing economies as they increase water consumption.

They need to know and understand the extent to which the world is water secure, the impacts of an increasingly water insecure world, and what future water scenarios for the world might be.

Examples used in the Student Book to study each key question are as follows:

**1 The geography of water supply: What is the geography of water supply and demand?**

- California's physical fresh water supplies, and their relationship to climate, river systems, and aquifers
- Water demand and stress in California
- The global water imbalance: stress, scarcity, water security
- The Millennium Development Goals and the importance of water in these
- Demand for water in the USA; conflicts within the Colorado Basin
- Water quality issues in Indonesia
- Links between global water insecurity, and poverty and wealth; focus on the USA and Indonesia

**2 The risks of water insecurity: What are the potential implications of an increasingly 'water insecure' world?**

- California: pollution of its supplies, over abstraction, salt-water incursion.
- Environmental issues in the Sacramento-San Joaquin river delta; salinity issues – the Salton Sea
- Water stress in the Middle East (Turkey and Israel) and North Africa (trade in virtual water in Kenya and Egypt)
- Turkey, Egypt, Kenya – agreements and treaties between areas; water abstraction on the Nile
- Turkey (agreements with Israel), Egypt (Toshka project), and Kenya (virtual water)
- Australia: water issues in the Murray-Darling Basin

**3 Water conflicts and the future: What are the possible conflicts and solutions to increasing demands for water?**

- Future water supply and demand in the Murray-Darling Basin, Australia
- Different approaches to water exploitation: privatisation (Bolivia) or public supplies (China's technological fix)
- Desalinisation in Israel; water transfer in the Middle East

## Chapter 3 Biodiversity under threat

This topic focuses on biodiversity as a key resource, its importance to human and ecological well-being, and the pressures placed on biodiversity by economic development.

Examples with which students should be familiar include global patterns of biodiversity, threatened biodiversity 'hotspots' such as the Daintree in Queensland, Australia, and ways in which biodiversity is valued differently by various interest groups. Focus should also be given to sustainable development in different parts of the world and ways in which biodiversity contributes to the Millennium Development Goals.

They need to know and understand the nature of biodiversity and biodiversity hotspots, threats to these, and prospects for sustainable development in future.

Examples used in the Student Book to study each key question are as follows:

### 1 Defining biodiversity: What is the nature and value of biodiversity?

- Ways of defining biodiversity: genetic, species, and ecosystem diversity
- The processes that determine biodiversity, and their relative importance, e.g. in the tundra
- Global patterns of biodiversity
- Threatened 'hotspots', e.g. the Daintree, Queensland, Australia
- The perceived value of biodiversity to different interest groups, e.g. Daintree, Alaska's ANWR

### 2 Biodiversity threats: What factors and processes threaten biodiversity?

- The Daintree in northern Queensland, Australia (a major study, used to knit several concepts together)
- Mangroves in south-east Asia (major study)
- Threats to Caucasus, southwest Australia; Atlantic Forest (South America)
- Deforestation in West Africa, Central America, Brazil, and Queensland
- The Nile perch (Lake Victoria); foxes and cats (southwest Australia)
- Eco-regions, e.g. the Daintree, the Sunda Shelf

### 3 Managing biodiversity: Can the threats to biodiversity be successfully managed?

- Sustainable yields and mangroves: shrimp farming in Thailand vs sustainable uses (e.g. Yadfon project in Thailand)
- Balancing conservation and development in the Daintree
- Players in Alaska's ANWR, the Daintree, and mangroves in South-East Asia; why they often conflict
- Costs and benefits of strategies, e.g. Millennium Development Goals, the Millennium Ecosystem Assessment (MEA) biosphere reserves
- Sustainable management in mangroves and the Daintree
- MEA scenarios
- Options for the Daintree, ANWR, and mangrove development

## Unit 3 Contested planet continued

### Chapter 4 Superpower geographies

This topic focuses on the world's superpowers and power – the way it develops over time in some countries and regions, and changing balances of power in the current world.

Examples with which students should be familiar include the world's current and past superpowers and their global influence, together with the rise of the new superpowers such as Russia, India, and China.

Students need to know and understand how superpowers can be identified politically and economically, their role in global influence and decision-making, and how they maintain power (e.g. the USA and its influence through organisations such as the World Bank), as well as ways in which global culture transmits via large transnational companies. Teachers and their students will also find it useful to cross-refer to Chapter 5 'Bridging the development gap' to help understand further the impacts of some of the world's superpowers.

Examples used in the Student Book to study each key question are as follows:

**1 Superpowers geographies: Who are the superpowers and how does their power develop over time?**

- Defining 'superpower' by size, population, resources, military and economic indicators and influence
- UK colonialism: the development of trade between UK and colonies, and its cultural impacts
- Theories that explain how and why colonialism made such an impact: modernism, Mackinder's Heartland theory, Capitalism, evangelical Christianity, Social Darwinism, Dependency, and development theory
- The emergence of the USA and USSR as superpowers
- China as an industrial giant, and the growth of the EU
- The collapse of colonialism and communism
- Economic influence: modernisation theory vs communism; global organisations with US influence – IMF, World Bank

**2 The role of superpowers: What impacts and influence do superpowers have?**

- European colonialism; legacies of colonialism (Ghana)
- Dependency theory and influence of trade (e.g. neo-colonialism, trade, aid, debt)
- Superpower influence through economic blocs (e.g. EU) or international groups (e.g. G8, IMF, NATO, UN, WTO, World Bank)
- How these maintain power, e.g. investment, military force
- Energy and resource influence: Russia as an energy superpower, China as a consumer of resources (e.g. Australian iron ore)
- Global culture: brand names, film, TV and music, publishing; US culture and companies (e.g. News International), India, and Bollywood

**3 Superpower futures: What are the implications of the continued rise of new superpowers?**

- The changing world order – the BRICs
- Resource demands by superpowers (e.g. China on Australia's iron ore reserves)
- Environmental impacts of economic growth, e.g. air pollution in Beijing
- The rise of China; Russian re-emergence as a superpower; its political and economic influence as an energy provider; its relations with neighbours and Europe

## Chapter 5 Bridging the development gap

This topic focuses on the development gap and its origins, its theoretical basis in terms of how development takes place, and how the development gap might be reduced.

Examples with which students should be familiar include current and historical processes by which the development gap arose, and the impacts it has had in particular countries, together with its consequences for the world's poorest people, and strategies by which the development gap might be reduced, and by whom.

Students need to know and understand development theory, so that they are aware of how the gap between the world's wealthiest and poorest countries has arisen, and therefore how it might be reduced. Teachers and their students will also find it useful to cross-refer to Chapter 4 'Superpower geographies' to help understand further how and why the development gap arose.

Examples used in the Student Book to study each key question are as follows:

### 1 The causes of the 'development gap': What is the nature of the 'development gap'? How has it arisen?

- Measuring development; indicators of development (e.g. economic, HDI)
- Theoretical background: how the development gap arose (core/periphery theory, dependency theory, cycles of poverty – vicious and virtuous cycles, colonialism and neo-colonialism, dependency theory)
- The emergence of the World Bank, IMF, WTO (see also Chapter 4 'Superpower geographies' on this)
- The G8 and the Millennium Development Goals, with their potential for reducing the gap
- Colonialism and dependency theory; the development of neo-colonialism; Uganda, Ghana

### 2 The consequences of the 'development gap': What are the implications of the 'development gap' at different scales for the world's poorest people?

- Uganda: standard of living and life chances
- The growth of debt in the 1970s/80s
- Structural Adjustment Packages; IMF conditions and the drive for privatisation
- The role of the World Bank/IMF in the development gap
- Resolving the debt crisis: debt cancellation in Uganda and Africa
- Bangalore: contrasts in development and emerging costs, e.g. traffic, housing, poverty
- Ethnic disparities in South Africa and apartheid
- The caste system in India/Bangalore
- Benefits and problems brought by Bangalore's 'new economy', e.g. urban sprawl, unequal wealth among different castes

### 3 Reducing the 'development gap': How might the development gap be reduced and by whom?

- Neo-liberalism (e.g. 'economic man') and 'free market' development; water privatisation in Ghana
- Modernisation and investment: big development projects, e.g. Pergau Dam; Ghana's Akosombo Dam and aluminium smelter
- NGO bottom-up approaches in Uganda (Barlonyo, Equatorial College School) and Moldova (Gura Bi Cului)
- Populism: land redistribution in Zimbabwe
- The Millennium Development Goals: how these are being met in Uganda and Bangladesh

## Unit 3 Contested planet continued

### Chapter 6 The technological fix?

This topic focuses on technology, and the extent to which it can manage and solve some of the issues facing the world today.

Examples with which students should be familiar include comparisons between technologies in countries at different levels of development, such as in farming or in social welfare (e.g. drug treatments). They should study examples of how governments, individuals, and private companies invest in technology for different reasons, and how national or regional problems (e.g. water supply) can be approached using technology of different kinds.

Students need to know and understand how access to technology is related to economic and social levels of development, leading to a world in which some can rely on technology to solve problems, while others lack access to it even at basic levels. They need to realise that technology has its costs as well as benefits, and that it varies between large-scale top-down mega-projects and small-scale intermediate approaches.

Examples used in the Student Book to study each key question are as follows:

**1 The geography of technology: Why is there inequality in access to technology?**

- Defining 'technology'
- High-speed rail in France and Africa
- Global distribution of internet and mobile phone technologies
- Access to technology in farming: global comparisons between GM foods, Sri Lanka (agro-technology), and Carfocial, Colombia
- Natural hazards and technology: the Thames Flood Barrier compared to flood prevention in Dhaka, Bangladesh
- Variations in treatment of HIV/AIDS globally

**2 Technology and development: How far does technology determine development and resource use?**

- Technology and development; drug treatments for HIV/AIDS
- Government sponsorship of research in higher education; patent laws
- Reducing the technology gap, e.g. mobile phones in Afghanistan; solar power in India and Pakistan; GM crop use in Latin America and Africa
- Technology sometimes comes at a price, e.g. GM crop technology and environmental issues; pollution and responses in western Europe

**3 Technology, environment and the future: What is the role of technology in the management of the contested planet?**

- Tigray, Ethiopia: high, intermediate and low technology solutions to water supply
- Technological fixes and the potential to solve climate change; renewable energy technologies and carbon capture technology
- Energy futures and technology in Slovakia
- Exploring a range of technological futures

## Unit 4 Geographical research

Unit 4 consists of six research Options: only **one** should be studied. The six Options cover a range of human and physical themes in Geography, as follows:

- Option 1: Tectonic activity and hazards
- Option 2: Cold environments – landscapes and change
- Option 3: Life on the margins: the food supply problem
- Option 4: The world of cultural diversity
- Option 5: Pollution and human health at risk
- Option 6: Consuming the rural landscape – leisure and tourism.

The six chapters that comprise Unit 4 in the Student Book are **sample studies**, and are designed to help your students select which Option to study. Students can (and should) use them in their research, but they do not form complete courses. They will help to get students thinking about research and what they may wish to study, for example in the post-AS period in June–July.

Each Option chapter in the Student Book is eight pages long and has the same format.

- The first page introduces the Option theme – what it's about, and what students should cover if they study the whole Option.
- There are six pages of sources, together with some background, providing a detailed insight into the sources. Therefore, Chapter 11 on 'Pollution and human health at risk' is about asbestos mining in Western Australia and its health impacts in Australia and elsewhere. Students are guided through the pages, but need to think about each source carefully.
- The final page provides activities, useful websites, and the kinds of exam questions that students could meet if they studied this Option.

Each Option chapter in the Student Book provides about 2–3 weeks worth of work, including time spent in following up some of the references on the final page. By adding in time to introduce the Option and time for students to present their findings, the material should be sufficient for 4–5 weeks of teaching.

## Unit 4 Geographical research continued

### Chapter 7 Tectonic activity and hazards

This topic focuses upon the wide range of natural hazards caused by plate tectonics.

Chapter 7 focuses on Montserrat, and the links between tectonic process, risk, vulnerability, and impact. Further research by students would lead them to consider ways in which people respond to hazard risk in different ways, depending upon their knowledge, technology, and financial resources.

The specification has four key questions which require that students learn about:

**1 Tectonic hazards and causes: What are tectonic hazards and what causes them?**

- Tectonic hazards and disasters, and what makes tectonic activity hazardous
- Event profile of hazards, including frequency, magnitude, duration, and areal extent
- The causes of tectonic hazards
- Tectonic activity associated with different types of plate margins

**2 Tectonic hazard – physical impacts: What impact does tectonic activity have on landscapes and why does this impact vary?**

- The varying impact of extrusive igneous activity
- The formation and morphology of different types of volcano and the characteristics of different types of eruption
- The varying impact of intrusive igneous activity, both major and minor
- The effects that earthquakes can have on landscapes

**3 Tectonic hazard – human impacts: What impacts do tectonic hazards have on people and how do these impacts vary?**

- Why people live in tectonically active areas
- The range of hazards associated with different types of tectonic activity
- The specific impacts of a range of tectonic hazards in countries at different stages of development
- Trends in frequency and impact over time

**4 Response to tectonic hazards: How do people cope with tectonic hazards, and what are the issues for the future?**

- Approaches by individuals and governments to coping with tectonic hazards
- Specific strategies involved in adjustment to hazards in locations at different stages of development
- The effectiveness of different approaches and methods of coping

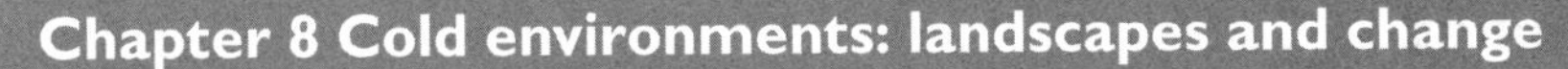

## Chapter 8 Cold environments: landscapes and change

This topic focuses upon cold environments which include glacial uplands, high latitude ice-bound regions, and periglacial areas.

Chapter 8 looks at:

- the Arctic, Norway, and landscapes of cold environments
- some of the conflicts that exist in these regions
- the links between processes occurring in the world's cold regions now and past Ice Ages in the UK.

Further research by students would lead them to understand links between the processes and landscapes that exist now, together with the fragility of such environments and the threats that they and the people who live there face.

The specification has four key questions, which require that students learn about:

**1 Defining and locating cold environments: What are cold environments and where are they found?**

- Cold, glacial, and periglacial environments, and the landscape systems operating there
- The varying nature of different cold environments
- Past and present distributions of cold environments, with reference to the British Isles

**2 Climatic processes and their causes: What climatic processes cause cold environments and what environmental conditions result?**

- The climatic causes of cold environments
- Long-term global climate change leads to changes in the distribution of cold environments
- The meteorological processes associated with cold climates
- The spatial and temporal relationships between glacial and periglacial environments

**3 Distinctive landforms and landscapes: How do geomorphological processes produce distinctive landscapes and landforms in cold environments?**

- Geomorphological processes in glacial environments and the landforms produced by these processes
- Geomorphological processes in periglacial environments and the landforms produced by these processes

**4 Challenges and opportunities: What challenges and opportunities exist in cold environments and what management issues might result from their use?**

- Past and present challenges, and opportunities of cold environments
- Attempts by people to take opportunities and overcome challenges
- The effectiveness of approaches to using and managing cold environments, and the conflicts that can exist

## Unit 4 Geographical research continued

### Chapter 9 Life on the margins: the food supply problem

This topic focuses on food insecurity, whereby some of the world's population live in regions where food production is a challenge. These are sometimes areas of traditional famine, but others are rapidly urbanising areas where food is scarce and malnutrition threatens.

The material in Chapter 9 is about food insecurity problems in countries such as India, Haiti, and Argentina. In fact, they reveal that food insecurity affects only some people in these countries; the Green Revolution in India has left some farmers very well off, whilst others are marginalised. Further research by students would lead them to understand links between these inequalities, together with the processes that have caused them.

The specification has four key questions, which require that students learn about:

**1 Global and local feast or famine: What are the characteristics of food supply and security?**

- Issues associated with food supply and security, e.g. food miles, famine, globalisation
- Environmental issues resulting from food production
- Why food supply varies spatially
- What life on the margins means to different people

**2 The complex causes of food supply inequalities: What has caused global inequalities in food supply and security?**

- The overlapping causes of famine and food surpluses, and classification of causes
- The role of population pressure in creating food insecurity
- The impacts of attempts to increase global food supply
- Who has been most affected by food insecurity, and why

**3 Desertification and life at the margin of survival: What is the role of desertification in threatening life at the margins?**

- Desertification, and its scale and impact
- The vulnerability of dryland ecosystems; why they are vulnerable to over-exploitation
- The relationship between food production and supply in desertified regions

**4 The role of management in food supply and security: How effective can management strategies be in sustaining life at the margins?**

- Techniques attempting to increase global food supply and security
- Why greater international efforts are increasingly needed
- Initiatives that have been most effective in sustaining life at the margins
- The role of sustainable strategies in food supply and security

## Chapter 10 The world of cultural diversity

This topic focuses on culture and how it varies spatially; some areas are homogenous, while others are hugely diverse. Large urban areas are often most diverse, reflected in their populations, services, and built environments.

Chapter 10 includes material about the Orang Asli people of Malaysia, and the ways in which their traditional culture is changing. Some of these changes result from interaction with urban areas, others result from the effects of globalisation, whilst more still are enforced upon them. Further research by students would lead them to understand how and why these changes are taking place, and how they pose threats to traditional cultures.

The specification has four key questions, which require that students learn about:

### 1 Defining culture and identifying its value: What is the nature and value of culture in terms of peoples and places?

- Defining culture, both human (ethnicities, beliefs, histories) and places (cultural landscapes)
- Human cultures and cultural landscapes continually change and evolve
- Some cultures and landscapes are more vulnerable than others from environmental, socio-economic, and political pressures
- The cultural diversity of people and places is valued differently by different players

### 2 The geography of culture: How and why does culture vary spatially?

- Some places are culturally more homogenous than others, e.g. Japan and Iceland compared to the UK
- Human cultural diversity is usually greatest in cities
- Attitudes towards human and landscape diversity can preserve diversity or move towards cultural homogeneity
- Global cultural imperialism affects cultural diversity and landscapes

### 3 The impact of globalisation on cultural diversity: How is globalisation impacting on culture?

- Views vary about the significance of globalisation in maintaining cultural diversity
- Global media corporations convey dominant cultural values and attitudes, and influence cultural globalisation
- Cultural globalisation often adapts locally, leading to hybrids of globalised fashion, music, and film
- Opinions vary about the impact of global consumerism on human and landscape culture

### 4 Cultural attitudes to the environment: How do cultural values impact on our relationship with the environment?

- Different cultures have different attitudes to the environment, landscape exploitation, and protection
- Anthropocentric cultural values are necessary to support and justify consumer cultures
- Conflicts between environmentalism and consumerism, and how to resolve these, e.g. the 'green' movement

## Unit 4 Geographical research continued

### Chapter 11 Pollution and human health at risk

This topic focuses upon human health, and the risks posed to health by economic development, either in the form of transmissible disease or environmental pollution. Pollution is a key risk especially in countries where rapid economic development takes precedence over environmental and health concerns.

Chapter 11 looks at asbestos mining in the Pilbara region of Western Australia. Focusing upon the former mining town of Wittenoom, the impacts of asbestos upon human health are explored, together with the ways in which asbestos exports to other countries spread the health risk elsewhere. Students will understand how and why asbestos poses a threat to human health, and the issues that arise from the use of asbestos in different countries.

The specification has four key questions, which require that students learn about:

**1 Defining the risks to human health: What are the health risks?**

- Defining human health risks
- Patterns of health risk at different scales (global, national, local)
- Health risk patterns over time
- How health affects both the quality of life and economic development

**2 The complex causes of health risk: What are the causes of health risks?**

- The complex causes of health risks
- The relationship between socio-economic status and health
- The links between some diseases and geographical features
- Models that help to understand health risk causes and patterns

**3 Pollution and health risk: What is the link between health risk and pollution?**

- The link between different pollution types and the health of societies
- The relative health risks associated with incidental and sustained pollution
- The link between pollution, economic development, and changing health risks
- The role of pollution fatigue to reducing health risk

**4 Managing the health risk: How can the impacts of health risk be managed?**

- The socio-economic and environmental impacts of health risk
- How health risk impacts have led to differing management strategies and policies
- The different agencies involved in health risk, especially international efforts
- Which health risks can be managed effectively and which cannot; and the role of sustainability

## Chapter 12 Consuming the rural landscape: leisure and tourism

This topic focuses on the structural shift taking place in rural areas, as they move from traditional production (e.g. farming, forestry) to consumption (e.g. towards tourism and leisure). The shift affects almost all rural regions, from accessible rural-urban fringes to the world's remote regions.

The material in Chapter 12 is about examples of changing rural landscapes in Peru and Hawaii, and the ways in which consumption puts pressure on often fragile rural landscapes, and represents a threat that requires careful management. Further research by students would lead them to understand the theoretical basis of these processes, and to see them as part of a broader economic and cultural shift.

The specification has four key questions, which require that students learn about:

### 1 The growth of leisure and tourism landscapes: What is the relationship between the growth of leisure and tourism and rural landscape use?

- The rise of leisure and tourism, and the spread of the pleasure periphery
- The range of rural landscapes sought for leisure and tourism activities
- The attitudes of different groups involved in this relationship, e.g. governments, businesses, communities
- How different leisure and tourism activities in rural landscapes may lead to conflicts

### 2 The significance and fragility of rural landscapes: What is the significance of some rural landscapes used for leisure and tourism?

- The physical significance and ecological value of some rural landscapes
- How rural settlements may be classed as fragile landscapes
- Theories explaining threats to rural landscapes, e.g. carrying capacity, resilience models
- Environmental quality measures, and their usefulness, in designating protected areas, e.g. National Parks

### 3 Impact on rural landscapes: What impact does leisure and tourism have on rural landscapes?

- The positive and negative impacts that leisure and tourism has on rural landscapes
- Ways in which impacts can change over time as levels of use vary
- The threats and opportunities posed in areas of differing economic development

### 4 Rural landscape management issues: How can rural landscapes used for leisure and tourism be managed?

- Whether rural landscapes should be managed or not
- Management strategies, e.g. preservation, conservation, sustainable management, and ecotourism
- Attitudes and strategies of different users of rural landscapes and conflicts that can exist between them
- The effectiveness of different approaches to managing rural environments

# Unit 3 Contested planet – synoptic investigation

Unit 3 Contested planet investigates the distribution of resources, plus the physical factors resulting in this distribution. The use and management of resources is a key issue for geographers in today's world. Many resources are finite and increasing consumption means that difficult decisions need to be made about their future use and management. Students will consider how resources are used, and also the problems and costs involved in obtaining them.

Three types of resource are considered in three topic areas:
- *Topic 1: Energy security*
- *Topic 2: Water conflicts*
- *Topic 3: Biodiversity under threat.*

Inequalities in global wealth, power, and influence are investigated in:
- *Topic 4: Superpower geographies*
- *Topic 5: Bridging the development gap.*

The role of technology in overcoming resource scarcity, income inequality, and environmental management is considered by investigating:
- *Topic 6: The technological fix?*

Part of students' learning will involve linking the content and concepts from Unit 3 with Unit 1 (Global challenges) and Unit 2 (Geographical investigations) in the AS course, in a synoptic investigation.

## Providing practice from an early stage

The synoptic element of A2 is important, and it is therefore advisable to give students experience of what to expect in this part of the examination, well before they encounter the real thing. To this end, the Student Book provides a synoptic enquiry within **every** topic in Unit 3, i.e. in Chapters 1–6. Generally, each enquiry occurs towards the end of each chapter, and is designed to:

- provide experience in interpreting source materials;
- develop skill in handling different viewpoints;
- encourage students to make links not just within each enquiry, but, as time progresses, between the different topics;
- give students experience of examination-style questions on these materials.

The themes covered by the six synoptic enquiries are:
- Chapter 1: Energy Security – Meeting future energy needs
- Chapter 2: Water Conflicts – Africa's water crisis
- Chapter 3: Biodiversity under Threat – Assessing the future of biodiversity
- Chapter 4: Superpower Geographies – Cultural superpowers and their influence
- Chapter 5: Bridging the Development Gap – Looking at aid and investment to fund development
- Chapter 6: The Technological Fix? – Energy security and independence

In addition, Chapter 13 provides a completed synoptic investigation on Dubai, which is designed to give a complete overview of Unit 3.

***Using the synoptic enquiries in Chapters 1–6***

Given the time allowance of 3–4 weeks for teaching each topic in Unit 3, you will need to allow time for synoptic practice. Remember that the idea of the synoptic units may be a new approach for many students. It is recommended that you therefore structure the earliest ones first, and gradually allow students more responsibility for reading and interpreting what is there as the course progresses.

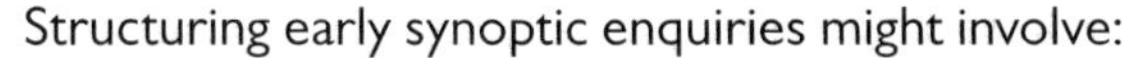

Structuring early synoptic enquiries might involve:

- Making sure the students know the purpose of the exercise – to draw the Unit together, and to provide experience of one of the major features of the exam.
- Reading and guiding students through the materials, whilst questioning them to ensure that they link the resources together.
- Ensuring they are clear about what issues are at stake in each synoptic enquiry.
- Class preparation of some of the questions given, followed by write-up in private study time.

Teaching the later synoptic enquiries might involve:

- Putting responsibility on to the students to read and digest what the materials are about.
- Discussion to ensure that they understand the materials, and questioning them about any links between each of the resources, and where they agree or conflict.
- Either giving a timed exercise in class to do the 70-minute full range of questions, or setting a timed task to be done in private study.

See pages 122–124 for information on using Chapter 13.

## Teaching the synoptic investigation

Four working weeks before the exam for Unit 3, Edexcel will issue a pre-release resource booklet. This is likely to contain 4–5 pages of resources about one of the six topics within Unit 3. It should usually be easy to identify which one! For example, the specimen materials on Edexcel's website are about GM foods, and therefore link to the Unit on 'The technological fix?' Once you have established this, you know that this topic will only be examined in Section B, and not Section A. So students can be told straightaway that they will not be required to revise this as such, but to integrate what they know and understand into their answers to the questions.

Note the following points:

- The resource booklet is intended to form the basis of student work in lessons for the weeks leading up to the exam. It is not expected that the booklet will be issued to students and then left to them; do ensure that you include time for this in your scheme of work.
- Lessons can be spent familiarising students with the materials, and the issues and concepts behind them. Four weeks is plenty of time to grasp the issue and absorb the detail.
- Students may not take the pre-release materials into the exam; a new set will be issued to them. Make sure they are familiar with this requirement.
- The booklet will always provide sufficient material for the exam questions – but will also always recommend extra websites for further research. Make sure that this is clear to students, and allow lesson time for those who may not have access to computers outside school or college.
- In the exam, the synoptic questions will be Question 6 in Section B. Students are recommended to spend 70 minutes on this question – longer than for the others on this exam. It carries 40 marks, which will normally be spread over three sub-questions – a, b and c.

## Preparing students for the synoptic investigation

To ensure your students succeed in the exams, make sure:

- that they are clear about which part of the specification is being assessed through the pre-release materials;

- that they read the materials thoroughly and can use evidence available in the resource materials – photos, diagrams, tables of statistics, etc. – to support their answers;
- that they use *all* the resources – they are included for a reason.

If your students are taking the Unit 3 exam in June, remember that you may not get the full four weeks' preparation time. Use your school or college website to provide support materials that help students to prepare for the exam. These should include:
- hyperlinks to the Edexcel website from which the resources can be downloaded;
- some basic comprehension exercises taking students through the materials to help them grasp the wider picture, but also the detail provided in some resources, and ensure knowledge of key terms;
- hyperlinks to websites recommended within the source materials, together with any others you have found;
- guidance on analysis, e.g. exploring the arguments for and against something, or analysing conflicting viewpoints, or environmental impacts;
- extended writing practice – don't try to question spot, but you can give students good practice at timed writing by suggesting how to spend the 70 minutes exam time, and how to ensure that they do not overshoot their time by writing too much;
- guidance on the focus for questions on the resources. These are likely to be:
  - What is the issue here? Why is this issue contentious?
  - an analysis of conflicting evidence in favour of, or against, a particular idea, or set of proposals;
  - assessing impact on people, economies, or environments;
  - projecting forward or evaluating, i.e. what could be done, or should be done, or assessing how successful something has proved.

## Preparing students for the exam

Every teacher has their own unique way of helping students to prepare for exams. The three points below are the most common reasons why students do, or do not do, well in examinations, so these especially apply to a tight time schedule like that in this exam.

1. **Plan.** Drum home to students that time spent planning is well spent. At the start of every question, encourage students to make a few pencil notes planning what to say. A few notes may be all that is needed. All the research into students who plan shows that they get more marks, because:
   - they stick to the point – having a plan stops them from going off-track;
   - they don't suffer from 'memory blanks' in which they forget what to say.
2. **Answering the question.** Give them practice in returning to a question to answer it. If the question is: 'Assess the human and environmental impacts of GM farming in Latin America', then simply listing impacts will only get them so far. If the impacts are classified into social, economic, and environmental, they will do better, as will be the case if they keep returning to the question to answer it.
3. **Timing.** Tell students to keep to a rigid time schedule – with 70 minutes for 40 marks, then about 10 minutes in total should be used for thinking and for planning, 50 minutes for writing and 10 minutes for reading, checking, and adding any other points that come to mind.

## Being synoptic!

The questions on the resource booklet in Section B are synoptic – that is, they are intended to draw out what students have learned across the whole of the AS and A2 courses. Using these materials, they are expected to make links between topics and

demonstrate prior knowledge. For example, the example of GM crops in the question above links to many topics within Unit 3, and elsewhere in the AS and A2 courses:

- biodiversity and its links to economic development;
- bridging the development gap – how technology in the developing world may bring benefits or problems;
- water and energy – the resource demands of GM crops.

To obtain maximum marks, students need to use examples in their writing that show what they have learned. For instance, in the question above on GM crops, reference could be made to examples where there are similar instances. Examples can be drawn from across the AS course, for example:

- how, in 'Re-branding places', examples were studied where appropriate investment in rural areas can help to develop local foodstuffs or organic foods without the use of GM technology, or where globalisation is linked to the spread of technology;
- how, in 'Bridging the development gap', the activities of TNCs and international organisations do not always benefit developing countries (with examples).

# Unit 4 Geographical research

Unit 4 Geographical research offers six Options for research, of which students choose one. These range from those with a physical geography focus, to those exploring environmental, social, and cultural geographies. They are designed to allow students to not just learn about new geographical ideas, but also to learn how to research independently (see page 33 for more information on research).

The Options on offer are:

- Option 1: Tectonic activity and hazards
- Option 2: Cold environments – landscapes and change
- Option 3: Life on the margins – the food supply problem
- Option 4: The world of cultural diversity
- Option 5: Pollution and human health at risk
- Option 6: Consuming the rural landscape – leisure and tourism

Chapters 7–12 in the Student Book are sample studies, which are designed to help students select which Option to study. Students can, and should, use these chapters in their research, but they do not form complete courses. For extra resources on this Option refer to *A2 Geography for Edexcel Activities & Planning OxBox CD-ROM*.

## Using the Student Book for Unit 4

The Student Book offers *sample studies* for Unit 4. The reasons for this are:

- Unit 4 is a research module. It would not be in the spirit of the Unit to provide coverage of all the material. It is not the kind of Unit that requires a 'one-stop-shop'.
- Nearly three-quarters of students that you teach will go to university, where there will be no 'one-stop-shop' textbooks. Unit 4 is excellent preparation for mature A2 students who will, just a few months later, be studying under their own steam.
- In the exam, marks are specifically awarded for evidence of research and citation from different sources. To provide a single source would counteract that, and would not do student potential any justice.
- With Unit 4, more than any other, it is vital that students have the most up-to-date research. For example, there are trends in the impact, prediction, and management of tectonic events that can alter thinking. The material available online in 2009 on tectonic hazards is quite different from that 10 years earlier, because of major events such as the 2004 tsunami which exposed vulnerable populations probably more than any previous event.

Therefore, the Student Book provides material which is primarily intended to provide students with an introduction to all six Options, so that they can make an informed choice. The material can, of course, form part of a student file.

However, there is considerable support material available on the *OxBox*. Unit 4 material there provides substantial guidance on web and published resources, including up-to-date websites. These can be uploaded on to your school or college intranet or VLE.

## Planning Unit 4

Unit 4 requires 40% of teaching time in the A2 year, since it is weighted at 40% of the assessment. Including a four to six week post-examination period after AS exams, this means about 40% of 30–32 weeks' teaching time, or about 13 weeks in total. When Unit 4 is taught depends upon several factors, particularly how many teachers are involved in teaching each class. Two models are suggested below: one where a teacher has sole responsibility for a class, and the second where that class is shared with one other member of staff.

***Single teacher***

| Post-AS period Year 1 | Term 1 Year 2 | Term 2 Year 2 | Term 3 Year 2 |
|---|---|---|---|
| • Introduce and begin teaching Unit 4 | • Unit 4 continues to early-mid November<br>• Unit 3 taught from early-mid November | • Unit 3 continues throughout term<br>• Unit 4 exam towards end of January | • Preparation of Unit 3 synoptic investigation |

***Two teachers, half teaching commitment each***

| Post-AS period Year 1 | Term 1 Year 2 | Term 2 Year 2 | Term 3 Year 2 |
|---|---|---|---|
| • Introduce and begin teaching Unit 4 | • Unit 4 continues throughout term | • Unit 4 taught until January exam<br>• Post-exam, picks up 1–2 units of Unit 3 | • Preparation of Unit 3 synoptic investigation |
| • Introduce and begin teaching Unit 3 | • Unit 3 continues throughout term | • Unit 3 continues throughout term | • Preparation of Unit 3 synoptic investigation |

## Teaching Unit 4

One of the most demanding skills in teaching a research module like Unit 4 is to manage students studying different Options in the same classroom, at the same time. Whilst some teachers are wary of such an approach, remember that many others have offered choices of Options to their students year-on-year in previous generations of Edexcel Geography specifications. They had to take the first plunge at some stage!

Factors that are likely to encourage you to offer a choice of Options include resource availability, and particularly access to computers. If you have access to computers for many or most lessons, then teaching becomes far more flexible and it is likely to encourage you to offer the Options in the true spirit of the specification. However, do remember the value of textbooks too, particularly in providing theoretical background that websites often ignore.

Remember, too, that the evidence from past Edexcel Geography specifications has shown that students who are given experience of research often gain higher marks than those who have been taught in a more controlled environment. Examiners award marks where they see evidence of research by the student (e.g. citing sources, or illustrating with examples and data), and it quickly becomes clear from a script whether or not a student has been taught research skills as well as content.

Below is a model by which you can teach Unit 4; you can use this either in a fully controlled environment with all students learning the same option, or in a situation where students have choice.

- In either case, the model of teaching and learning suggested is that of the teacher as a tutor and guide rather than a provider of resources and content. To differentiate how this might mean different styles from the rest of the AS/A2 course, the roles of teacher and student are identified separately.

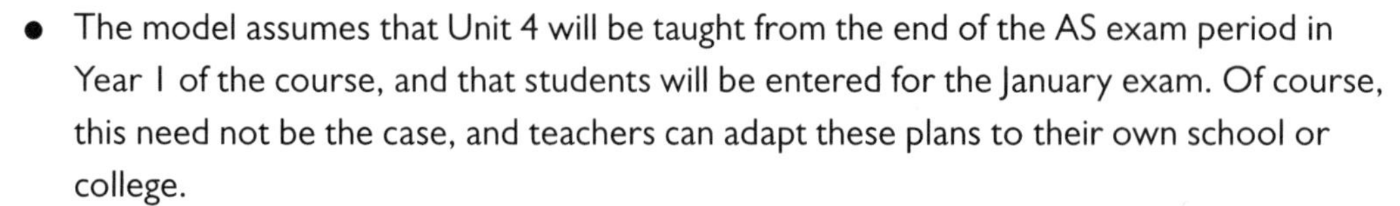

- The model assumes that Unit 4 will be taught from the end of the AS exam period in Year 1 of the course, and that students will be entered for the January exam. Of course, this need not be the case, and teachers can adapt these plans to their own school or college.
- Each model assumes that the entire teaching week is devoted to Unit 4, and that Unit 3 will be taught at some other time. If you are sharing the teaching of the course with another member of staff, and therefore only getting half a week, you will need to double the number of weeks allowed for each stage as shown below.

Remember that this is a research module; whether you elect to give students choice is less important than giving them time and opportunity to research. Don't expect them to do this outside class; 40% of students nationally have no computer at home, and rely mainly upon school for whatever access they can get. So build in activities that are research-focused into the course. The spirit of Unit 4 is that students should research, and it is enforced through the exam mark scheme. Build a course that offers plenty of opportunity for students to research, feed back, and present their findings. It will result in a wide diversity of case-study examples that will enhance student understanding of the concepts and the Option being studied.

## A teaching and learning model for Unit 4

***Post-AS period Year 1***

Note that Weeks 1–4 apply only where choice is being offered by the teacher.

***Weeks 1–3***

- **Teacher** introduces students to Unit 4 and the Options available, sets up choices, and divides students into like-minded groups. Teacher supervises groups, tutoring and intervening with ideas where necessary. Teacher sets an agenda for how the group should work in this, very different, teaching unit. Emphasise the need for self-motivation, group collaboration, and good time management.
- **Students** review the Options and make a short list of which one or two they think they will enjoy most. They work in small groups of two or three on the sample materials for Unit 4 in the Student Book (Chapters 7–12). This should include some investigation of further sources on the last page of each chapter, including some of the music, films, or books listed.

***Week 4***

- **Teacher** outlines what is required over the next one to two weeks, and sets a requirement for student presentations, tutoring and intervening with ideas where necessary.
- **Students** begin presentations on the sample materials they have studied, to be given to other students. They should focus on:
  - what their chosen Option is about;
  - what its most interesting points seem to be;
  - case-study examples from the sample materials to show other students what they have studied;
  - what other kinds of topics other students might study for this Option (refer to the last page of each chapter).

***Weeks 5–6***

- **Students** begin work on investigating the first enquiry questions for their Option.
- **Teacher** provides the framework for students to complete when listening to other presentations; supervises and assists with presentations where appropriate;

challenges and manages discussion post-presentation; gives positive feedback to each group, together with structured peer evaluation of each presentation; sets small research tasks, for example:

| Option | Suggested activity |
|---|---|
| Tectonic activity and hazards | Investigating the types of hazard generated by tectonic events |
| Cold environments – landscapes and change | Investigating the range of different cold environments: glacial, periglacial, mountainous, etc. |
| Life on the margins – the food supply problem | Investigating the range of food supply problems, e.g. food miles, famine, globalisation of food tastes, under- and over-nutrition |
| The world of cultural diversity | Investigating the range of human cultures, their shared characteristics, attitudes, beliefs and values, and the links between culture and landscape |
| Pollution and human health at risk | Investigating a range of health risks from short-term to chronic, and their geographical patterns |
| Consuming the rural landscape – leisure and tourism | Investigating the wide range of different uses that leisure and tourism activities make of rural landscapes |

These can be presented to students in the same group. Organise a rota so that you can sit in on their presentations. Place the presentations on the school or college website for future revision and reference. Provides students with a research brief for the summer holiday.

***Term I Year 2***
***Week I***

- **Students** are grouped into those selecting the same Option.
- **Teacher** outlines Unit 4 to recap – what it is about, and the assessment format – and outlines what a research module is like. Sets an increasingly broad research exercise, covering wider aspects of the specification for each Option, for example:

| Option | Suggested activity |
|---|---|
| Tectonic activity and hazards | Investigating the impacts of tectonic activity upon landscapes |
| Cold environments – landscapes and change | Investigating the impacts of glacial processes upon landscapes |
| Life on the margins – the food supply problem | Investigating the advantages and disadvantages of GM foods set against organic farming as ways of increasing food production |
| The world of cultural diversity | Investigating the effects of cultural imperialism and the influence of TNCs such as MacDonald's or Disney in two culturally different countries |
| Pollution and human health at risk | Investigating the influence of socio-economic status upon health care in different urban areas |

| Consuming the rural landscape – leisure and tourism | Assessing a range of threats to rural landscapes in two or three different areas |
|---|---|

***Weeks 2–7***

- **Students** work together in lessons. Occasional presentations given at teacher direction by students studying the same option.
- **Teacher** starts each lesson with a review of progress; organises a rota to see each group and sit in on presentations; places student presentations on the school or college website for future revision and reference; provides students with research guidance; sets sample essays with increasingly broad titles.

***Weeks 8–9***

- **Teacher** begins to prepare students for what to expect for the exam, sets timed essays, and teaches guidance about what makes a good essay.

## Preparing for the exam

Four working weeks before the exam, research focus material is pre-released to candidates as advance information via the Edexcel website (www.edexcel.com).

This advance information will consist of the release of the **two** themes that will be assessed in that examination. There are four themes in total for each Option, so the purpose is to limit student research in the last four working weeks, thereby intensifying focus and knowledge in depth.

Teacher and students prepare materials from research files of relevance to the focus. Students step-up the number of timed and practice essays.

## The exam

Candidates are given a list of questions based on the six Options. They select and answer the single question that relates to the Option they have studied; there is no choice within the Option they have studied. They are required to write a long essay in which they demonstrate and synthesise the results of their research. Each question is out of 70 marks. Candidates will not be able to take any pre-released or research materials into the exam.

# Integrating research into the course

This section is designed to help you integrate research into Unit 3 Contested planet, as well as Unit 4 Geographical research, where it is a requirement of the course. Research at A2 for the Edexcel specification is important for two reasons.

- Students can develop **responsibility for their own learning**, helping to prepare them for university and employment. This is particularly true where they are encouraged to work in small groups, share and discuss ideas, and solve problems.
- **Content can be updated**. The global financial crisis in 2008–9 is likely to have several major impacts which relate to the themes in both Units 3 and 4. Research, using some of the guided exercises in the Student Book, will allow both teachers and students to assess the impacts of the financial crisis on, for example, 'Superpower geographies' or 'Bridging the development gap'.

## The importance of research

With their contemporary feel, the six topics in Unit 3 lend themselves to further research by students. Concepts are the foundation of the specification, for example: What characteristics make a country a superpower? Because it is so contemporary, some detail is likely to change over time. For example, during the writing of the Student Book, China moved from fifth to fourth in the rank of the world's largest economies. By the time the first students sit their A2 exams for this specification, it is likely to be third. Research by students will enable them to have the most up-to-date information to ensure that they keep abreast of change.

Therefore, take note of websites and research exercises in the Student Book, which will provide you and your students with data. The CIA Factbook (key 'CIA Factbook' into Google) is excellent for up-to-date data, and is probably easier to negotiate than the World Bank (worldbank.org).

## Creating successful research exercises

The Student Book provides several structured enquiry exercises to help students research, to provide a focus for that research, and to provide a forum for feedback. These exercises have been written by teachers who use exercises exactly like these – and they work! Always remember that, as a teacher, you can never know everything there is to know in an ever-changing world. What you can do is have a framework of ideas in your mind, into which students can slot the content that they have researched. That, after all, is what they will be expected to do at university or in employment. Never feel that you have to be an expert on data or facts which will change; your job is to help students understand the general trend and framework into which these fit.

## The keys to successful research

The success of good research does not lie in just sending students to find out. The value of research lies within learning purpose and structure. To get the best from research exercises follow these simple procedures.

- Take account of **student experience in researching**. If students have always been given notes and headings previously, or have that kind of experience in other subjects, you will need to ease them into research gradually and may even need to sell it to them. Make the first few research exercises short, limited in scope, and very focused. It is disconcerting for you and for students if the first few exercises are unsuccessful. As they gain research experience, you can extend the tasks and develop wider boundaries for their work.

- Make it absolutely clear **where the research fits** within the topics they are studying, so that they see its relevance, and therefore have a 'need to know'.
- Give a clear **structure** to the enquiry, so that students know upon what they should be focusing. For example, focus the research upon a question or problem. Instead of researching 'Uganda and the Millennium Development Goals (MDGs)', you could focus on 'How well is Uganda progressing in the MDGs?'
- **Split the research exercise** up between different students or groups of students, for example, finding out about the success of the MDGs among different countries. In this way, students know that their research fits a framework, can report back on their research, and share ideas. Then the effect of the research is cumulative.
- Make sure every student in a working group has a **discrete role**. Don't expect a group of three students to collect data all on the same thing. All that will happen is that two others may let one person do all the work!
- **Give guidance**, for example, give one or two websites which will provide at least some of the research material. Early success is crucial.
- **Define clear boundaries** to the research, for example researching economic data about country X, rather than just 'data'.
- **Limit the length of presentation** or findings required. If you want no more than four or five slides of a PowerPoint, or two sides of a handout, say so.
- Encourage students to **read, edit, and filter** their findings, so that they don't just 'find, copy, paste, present'. Develop their research reading skills – see Developing reading skills' below for more guidance on this.
- Give a clear **structure for feedback** in class. Make it clear who will feed back on what, in what order, and with what purpose. You can vary this as students develop confidence and skill; you'll find that they will quickly define their own structure, and be able to act more freely.
- Make sure **every group** has a chance to feed back; there is little more de-motivating than two or three groups giving all the feedback and others feeling of little worth. If the feedback will take more than one lesson, say so at the outset, and make it clear that all groups will get a chance.
- Give all class members **something to do** while presentations of research are taking place, for example, if you want to compare success in achieving the MDGs in different countries, create a table listing all the countries, their successes so far and targets still to be achieved. Students can then complete this as they hear each presentation.
- **Make available the results** of all research afterwards – for example, via a combined PowerPoint or on the school website – thus making these available for student revision.
- **Debrief** the exercise after. Do this by:
  - drawing together points after each presentation;
  - making links between different findings as they occur;
  - drawing the exercise together by helping students to compare different findings, analyse, and draw conclusions
  - evaluating the exercise in terms of its success, for example, good websites used, the validity of findings, where further enquiries might lead, how a future enquiry in two or three years' time might show different findings, etc.
  - providing positive feedback and points to work on for every group.
- Give a **meaningful follow up** exercise so that the research does not 'stand alone'. For example, in 'Superpower geographies', Unit 4.2 activity 3 in the Student Book involves students researching criteria to help decide which countries can be deemed

'superpowers'. You could follow this with a written conclusion about which countries can be considered superpowers, with justification.

## Developing reading skills

Successful research means successful location and identification of material. All too often, students will quickly find a single appropriate website, highlight it, copy it, and paste it into a document, presenting it as their own work. Aside from issues of plagiarism, they should be encouraged to adopt much better practice, because this will lead to greatest success in the exam. So:

- from day one, encourage students to log, refer to, and acknowledge **sources**. In this way, you can assist their learning where they misquote or misunderstand data or sources. Let them know that they will be expected to quote organisations and sources in the exam, and that this will gain them marks;
- to help this process, make **class time** available for research so that you can model and promote good practice. Unit 4, after all, is a research module, so class time spent researching and processing information should be one of the major activities!

Reading material is a further issue. Reports and several relevant internet sources are far too large and unwieldy to expect students to read them all. Help them by showing that the following process can help, once they have located a source website.

- **Scan** – Does the website suit the focus? Browse fairly quickly and see what it contains by reviewing headings, charts, tables, and any illustrations.
- **Skim** read – Take what might be relevant sources and skim over them, skim reading a paragraph or two. This will give a better idea of whether the website is relevant.
- **Review** the rest of the document, for example, its sub-headings, illustrations, tables of data.
- Continuous, **close reading** should follow only once the source is identified as useful.
- At this stage, open up a Word or similar text document. They can certainly copy parts of a document into this – but encourage them to edit and make notes on what the text is about. A good exercise is **summarising**; if they paste 300 words into Word, can they then edit down to 50 words?

Working in this way will help students to do something with their research, by filtering material and getting rid of the chaff. It will encourage them to edit, and reflect critically on sources. This approach is likely to prove so much more productive than just simply saying 'make notes on…' which remains a dominant activity in post-16 teaching.

# Energy security

## Chapter outline

Use this chapter outline and the introductory page of the chapter in the Student Book to give students a mental roadmap for the chapter.

**1.1 Energy security in the USA** Energy security issues in the USA as a whole, and California in particular

**1.2 The energy mix** The environmental impacts of different energy sources and the global distribution of the main energy sources

**1.3 Access to energy** How a range of factors affect the availability of, and access to, energy

**1.4 China: the new economic giant** How China is coping with its enormous demand for energy

**1.5 Energy connections and geopolitics** The complexity of energy pathways, and the economic and political risks if energy supplies are disrupted

**1.6 The race for new resources** How increasing energy insecurity is leading to the exploration of technically difficult and environmentally sensitive areas in the search for new resources

**1.7 Energy supply players** The economic and political power of OPEC and energy companies in the USA

**1.8 Is there enough energy?** There is uncertainty over the global energy supply in terms of demand, reserves, and peak oil and gas

**1.9 Business as usual** Reliance on fossil fuels – the costs, geopolitical tension, and potential for conflict

**1.10 Meeting future energy needs** Approaches that we can adopt to meet our future energy needs and increase our energy security (Synoptic unit)

## About the topic

- This topic has two underlying themes – energy supplies and the conflicts between those who have energy supplies and those who do not.
- Global energy supplies are not evenly distributed. Some areas have energy surpluses and are energy secure, while others have too little and are insecure.
- Physical factors usually determine supplies, whilst human factors determine consumption, and the pathways between those who have energy and those who do not.
- Growing energy demands do not match supply; resources are often located far from where they are consumed, and the pathways between these can be insecure. The potential for conflict is high.
- Increased demand is leading to demands for the exploitation of energy reserves in sensitive areas.

## About the chapter

- This chapter looks at energy security issues in the USA and increasing demand for energy in China coupled with energy insecurity.
- It investigates: the environmental impacts of different energy sources; the global distribution of energy sources; the availability of, and access to, energy; energy pathways; and the geopolitics of energy supplies.
- It examines how energy insecurity is leading to the search for new resources, such as oil under the Arctic Ocean and the development of Canada's tar sands.
- It looks at energy supply players – the economic and political power of OPEC and energy companies in the USA, and the issue of peak oil and gas.
- It concludes by investigating how we can meet our future energy needs – by continuing with 'business as usual', and looking at alternative approaches.

## Key vocabulary

There is no set list of words in the specification that students must know. However, examiners will use some or all of the following words in the examinations, and would expect students to know them and use them in their answers.

cartel
energy dependency
energy pathways
energy poverty
energy security
energy surplus
frontier hydrocarbons
geopolitics

low-carbon standard
OPEC
peak oil
production hotspot
security premium
strategic
tar sands
transit state

The glossary at the end of this book contains many of these words and phrases. For students, the key word boxes in the chapter or the glossary at the end of the Student Book will help them with the meanings of all.

# 1.1 Energy security in the USA

## The unit in brief

The chapter on 'Energy security' begins with a five-page unit which looks at energy security issues in the USA as a whole, and California in particular. The USA suffers from energy insecurity and faces an energy crisis as a result of its enormous energy consumption, its reliance on imports, and the rising price of oil.

California's energy crisis in 2000/2001 was characterised by price instability and blackouts which affected millions of people. California has since begun to lead the way in efforts to conserve energy supplies, look for alternative sources of energy, and reduce greenhouse gas emissions.

## Key ideas

- The USA suffers energy insecurity as a result of its energy consumption, its reliance on imports, and the rising price of oil.
- Energy security is a global problem as reserves of fossil fuels are beginning to run out, sources of energy are unevenly distributed, and demand for energy is increasing.
- There is a range of factors which affect energy security.
- California's energy crisis of 2000/2001 was characterised by price instability and major blackouts.
- California has begun to lead the way in energy conservation, looking for alternative energy sources and attempting to reduce greenhouse gas emissions.

## Unit outcomes

By the end of this unit most students should be able to:

- draw a spider diagram to identify the factors threatening the USA's energy security and how its energy security could be improved;
- explain why energy security is a global problem;
- define the term 'energy security' and describe the factors which affect it;
- list the factors which led to the Californian blackouts;
- describe the changes in California since the blackouts of 2000/2001.

## Ideas for a starter

1. Show the photo on page 6 of the Student Book on the whiteboard. Dramatic oil price rises in 2008 pushed up the prices of petrol and diesel. The USA had always expected low-cost energy, but 2008 showed that those days were over. Ask: What has the price of petrol or diesel got to do with geography?
2. Ask students to read out the two speech bubbles at the top of page 6 of the Student Book. The USA was facing an energy crisis – but why? Ask students to come up with ideas and record them. (You could return to this at the end of the unit to see if students' ideas were along the right lines.)

## Ideas for plenaries

1. Begin to build a dictionary of key terms and their definitions for this chapter. Start with the term 'energy security'.
2. Use the 'What do you think?' on page 10 of the Student Book as a plenary to get students thinking more about the control of energy generation and energy supply. They will be meeting these issues again later in the chapter.
3. Question time. Ask students to think back over the unit and write down three questions related to what they have learned. Ask individual students to feed back.

# 1.2 The energy mix

## The unit in brief

In this 3-page unit students investigate energy classification, the environmental costs or impacts of different energy sources, and the global distribution of the main energy sources.

The world's main energy sources are distributed unevenly. The distribution of renewable energy sources also varies globally, but many parts of the world should be able to adopt one or other type of renewable energy.

## Key ideas

- Energy sources can be classified as non-renewable, renewable, and recyclable.
- Different sources of energy have different environmental costs associated with their production and use.
- The world's main energy sources are distributed unevenly.
- The distribution of renewable energy sources varies globally.
- Many parts of the world should be able to adopt at least one renewable energy source.

## Unit outcomes

By the end of this unit most students should be able to:

- classify a range of energy sources as non-renewable, renewable, and recyclable;
- decide which types of energy have the greatest and the least environmental impact;
- describe the distribution of the world's main energy sources;
- suggest how different parts of the world could adopt different sources of renewable energy.

## Ideas for a starter

1. Go around the class and ask students to name different energy sources. Then ask them to suggest the environmental impacts they have. Compare their responses with the table on page 11 of the Student Book.
2. Show either of the two maps on page 13 of the Student Book on the whiteboard minus either the key or the caption information. Ask: What does the map show? Reveal the key or give the information in the caption for the solar energy map. How does the global distribution of these energy resources determine their use?

## Ideas for plenaries

1. A quick quiz. Ask students questions based on the tables of energy reserves on page 12 of the Student Book. For example: Which country has the largest reserves of oil? Which has the largest reserves of natural gas? What does Australia have the most of?
2. You could use the 'What do you think?' on page 13 of the Student Book as a plenary. Are we ever likely to end our dependence on fossil fuels?
3. Ask students to summarise what they have learned in this unit in less than 40 words.

# 1.3 Access to energy

## The unit in brief

This is a 3-page unit in which students learn how a range of factors affects the availability of, and access to, energy. Access to energy depends on physical factors, public perception, technology, and cost.

Energy sources are consumed unequally around the world, with the USA heading the league for oil and gas consumption, whilst China consumes the largest volume of coal. Whilst some countries have an energy surplus, the unequal distribution of energy and lack of access to energy resources means that some countries, or areas, suffer from energy poverty.

## Key ideas

- Access to energy depends on physical factors, cost, technology, and public perception.
- Plans to build a new generation of nuclear power stations are intended to improve Britain's energy security, but many people are concerned about the plans.
- Energy sources are consumed unequally.
- The unequal distribution of energy, and lack of access to energy resources, means that some countries, or areas, suffer from energy poverty.

## Unit outcomes

By the end of this unit most students should be able to:

- rank the factors determining access to energy, and justify the rankings;
- explain how building new power stations will improve Britain's energy security, and why people are concerned;
- rank the consumption of oil, gas, and coal for a range of countries;
- give an example of how energy poverty can be addressed.

## Ideas for a starter

1. Find a photo of people protesting about E.ON's plans to build new coal-fired power stations at Kingsnorth, Kent. E.ON claims that the new power stations will be 20% cleaner than the existing one. Ask: Should E.ON be allowed to build the new power stations?
2. Ask: Who is in favour of nuclear power, and why? Who is against nuclear power, and why?
3. Ask individual students to read out the first three bullet points at the top of page 16 of the Student Book. Can students think of a term which describes these facts? (You are looking for the term 'energy poverty'.)

## Ideas for plenaries

1. If you used starter **2**, you could continue the debate by asking who is in favour of/against wind farms, tidal barrages, dams, and so on. Then continue with the 'What do you think'? on page 16 of the Student Book. If people object to virtually every method of electricity generation, where should we get our energy from?
2. Use 'Over to you' activity **1** on page 16 of the Student Book as a plenary.
3. Get students to add the term 'energy poverty' to their dictionary of key terms for this chapter.

# 1.4 China: the new economic giant

## The unit in brief

This 5-page unit is a case study of China. It investigates how China's demand for, and consumption of, energy has grown – fuelled by economic growth and the demands of new industry, as well as rapid urbanisation and growing car ownership.

The unit looks at China's reliance on coal for 70% of its electricity generation (it is self-sufficient in coal) and its plans to increase its output from HEP. Although China has the world's third largest reserves of coal, it suffers from energy insecurity. The unit ends by looking at how China is trying to improve its energy security.

## Key ideas

- China's rapid increase in energy consumption and demand has been fuelled by economic growth, the demands of new industry, rapid urbanisation, and car ownership.
- China is the world's second largest energy consumer (and is set to become the world's leading source of greenhouse gas emissions).
- China relies heavily on coal for electricity generation (70%) and plans to increase its output of HEP (currently 16% of energy production).
- China's energy insecurity matters because of the country's sheer size.
- Energy has become a national security issue in China.

## Unit outcomes

By the end of this unit most students should be able to:
- account for China's rapid increase in energy demand and consumption;
- understand that although China is the world's second largest energy consumer, its per capita demand remains relatively small;
- identify the issues raised by China's main methods of electricity generation;
- create spider diagrams to show the reasons for China's energy insecurity, and the steps that China is taking to tackle its energy insecurity.

## Ideas for a starter

1. Show the cartoon on page 17 of the Student Book on the whiteboard. Ask students for their views on the cartoon.
2. Tell students that China's huge demand for electricity means that it is building an average of three new coal-fired power stations a week. In 2006, China added 102 gigawatts of generating capacity to its grid – as much as the entire capacity of France. Ask: Why is demand for electricity rising so rapidly in China?

## Ideas for plenaries

1. With books closed, ask students to define the term 'energy dependency' and give an example of a highly energy-dependent country. Then get students to add the term to their dictionary of key terms for this chapter.
2. China's view on energy could be summed up by this statement made by Deng Xiaoping: 'I don't care if it's a white cat or a black cat. It's a good cat so long as it catches mice.' Use this as the focus for a discussion on China's energy policy.
3. Use 'Over to you' activity 1 on page 21 of the Student Book as a plenary.

# 1.5 Energy connections and geopolitics

## The unit in brief

The first three pages of this unit focus on energy pathways and provide a case study of the East Siberia-Pacific Ocean (ESPO) oil pipeline. It looks at the geopolitics involved between Russia, China, and Japan as the latter two compete for access to Russia's oil.

The final three pages look at the economic and political risks if energy supplies are disrupted. It focuses on Russia and Gazprom (the world's largest gas supply company). Russia cut off gas supplies to Ukraine in 2006 and 2008, which alarmed many European countries dependent on Russian gas.

## Key ideas

- The ESPO oil pipeline is an energy pathway that will link Russia with China, Japan, and any other Pacific countries which import Russia's oil.
- The term 'energy pathways' refers to the flow of energy from a producer and the means by which it reaches the consumer.
- China and Japan have been competing for access to Russia's oil and the ESPO pipeline project.
- Gazprom is the world's largest gas supply company, providing 25% of the EU's natural gas (with 80% of exports to Western Europe crossing Ukraine).
- Russia has been accused of using energy as a political and economic tool.
- Europe's dependence on Russian gas raises concerns about its energy security.

## Unit outcomes

By the end of this unit most students should be able to:
- outline the problems that the ESPO oil pipeline has faced;
- define the terms 'energy pathway', 'geopolitics', and 'transit state';
- identify the reasons why China and Japan want a share of Russia's oil;
- give five facts which underline Gazprom's importance as a gas supplier;
- explain why Gazprom's close links to the Russian government could create problems for countries dependent on Russian gas;
- assess how far Europe is right to be concerned about its energy security.

## Ideas for a starter

1. Show the photo of the Amur leopard from page 22 of the Student Book on the whiteboard. Tell students that these leopards live near Vladivostok and there are only 30–40 left in the wild. Ask: What does this animal have to do with energy security?
2. Ask students to explain the term 'geopolitics'.
3. Show the two photos on page 25 of the Student Book and read out the article about Gazprom to introduce the section on energy control and disruption.

## Ideas for plenaries

1. Use 'Over to you' activity **2** or the 'What do you think?' on page 27 of the Student Book for a class debate.
2. Get students to add the three key terms from this unit ('energy pathways', 'geopolitics', and 'transit state') to their dictionary of key terms for this chapter.
3. Give students five minutes to write a radio bulletin on the economic and political risks of the disruption of energy supplies.

# 1.6 The race for new resources

## The unit in brief

This 4-page unit investigates how increasing energy insecurity is leading to the exploration of technically difficult and environmentally sensitive areas in the search for new resources. It begins with Russia's flag planting below the North Pole in 2007 – in doing so Russia symbolically claimed the rights to the seabed and its resources. But eight countries form part of the Arctic Region and many of them have their eyes set on the vast energy and mineral deposits located there.

The unit also investigates Canada's tar sands. Tar sands provide an alternative source of oil when conventional sources are unavailable, but there are enormous costs associated with their exploitation.

## Key ideas

- It is estimated that the Arctic Region contains 25% of the world's unexploited oil and gas reserves (as well as diamonds, coal, and other minerals).
- Many of the eight countries that form the Arctic Region want a share of the energy and mineral deposits.
- A UN panel will decide about control of the Arctic Region by 2020, but in the meantime Arctic states battle on.
- The largest reserves of tar sands are found in Canada and Venezuela.
- Canada's reserves of tar sands could be second only to Saudi Arabia's reserves of conventional oil.
- There are costs and benefits of exploiting tar sands.

## Unit outcomes

By the end of this unit most students should be able to:

- identify the reasons why countries are racing to exploit energy resources in technically difficult and environmentally sensitive areas;
- complete a table to assess the economic, social, and environmental costs and benefits of drilling for oil in the Arctic Region;
- complete a conflict matrix for the players involved in mining tar sands in Canada and describe what it shows about exploiting the tar sands.

## Ideas for a starter

1. Show a photo of pristine Canadian forest on the whiteboard. Ask students to imagine what this area would be like if open-cast mining was allowed to take place there. Then show them the photo at the foot of page 30 of the Student Book. Show it as large as possible on the whiteboard – the destruction is total – all in the name of oil.
2. Show the photo on page 35 of the Student Book of the oil pipeline snaking across the Alaskan landscape and use the paragraph on the environmental view on page 29 to introduce the issue of the race for new energy resources. A student could read out Ben Stewart's statement.

## Ideas for plenaries

1. Ask students to write down as many words as they can relating to this unit. Get them to add any key terms to their dictionary for 'Energy security'.
2. Use the two speech bubbles at the foot of the text panels on page 31 of the Student Book to open up a debate on the costs and benefits of exploiting tar sands.
3. If you did not use starter **2**, ask students how far they agree with Ben Stewart's statement on page 29 of the Student Book.

# 1.7 Energy supply players

## The unit in brief

This is a 4-page unit looking at the economic and political power of OPEC and energy companies in the USA. OPEC is a major player in the energy supply business and has the ability to increase production if demand for oil rises (to prevent sharp rises in the price) and similarly reduce production if demand falls (to maintain the price). A Background box explains how the price of oil is decided.

The role of oil and gas companies in terms of political funding in the USA is also examined. The unit looks at the influence that oil and gas companies can exert and also at examples of how politicians work in the oil companies' interests.

## Key ideas

- OPEC is a permanent intergovernmental organisation, formed in 1960 and consisting of oil producing and exporting countries.
- OPEC has nearly 78% of the world's total oil reserves, and produces 45% of the world's crude oil and 18% of its natural gas.
- The oil and gas industry has a long history of influence on the US government.
- Since 1990, 75% of political funding from oil and gas companies in the USA has gone to the Republicans.
- Politicians in the USA work in the oil companies' interests.

## Unit outcomes

By the end of this unit most students should be able to:

- state OPEC's objective;
- use evidence to explain OPEC's importance in terms of energy supply;
- understand what determines the price of oil;
- give examples of how the oil and gas companies exert their influence on the US government;
- give examples of how political parties in the USA work in the oil companies' interests;
- list the advantages and disadvantages of cartels for producers and consumers.

## Ideas for a starter

1. Ask students to explain what a cartel is and to give some examples.
2. Show the table of where OPEC's oil went in 2006 from page 33 of the Student Book on the whiteboard. Ask: What does this tell us about OPEC and its member countries?
3. Show the table from page 34 of the Student Book on the whiteboard. Ask students what they think this is about. Why is it important?

## Ideas for plenaries

1. Get students to re-read the Background box on the price of oil on page 33 of the Student Book. Then use the 'What do you think?' on page 35 as a plenary.
2. You could use 'Over to you' activity **2** on page 35 as a plenary.
3. Show the photo of the Alaskan landscape on page 35 on the whiteboard. Use it as the catalyst for a debate on the influence that oil and gas companies have on politicians.

# 1.8 Is there enough energy?

## The unit in brief

In this 4-page unit students learn that there is uncertainty over the global energy supply in terms of demand, reserves, and peak oil and gas.

Demand for oil and other sources of energy continues to increase, and although China and India's energy use was predicted to double between 2005 and 2030, the West is still expected to dominate consumption for the foreseeable future. While it is possible to increase oil production to meet increasing demand, this is expensive and takes time. However, oil is a finite resource and production in some countries has already peaked, and is now in decline.

## Key ideas

- Demand for oil and other sources of energy continues to rise.
- The West is expected to dominate energy consumption for the foreseeable future.
- Saudi Arabia has the largest reserves of oil and Russia has the largest gas reserves.
- Increasing oil production is expensive and takes time, but the amount of exploitable new oil discoveries has fallen since the 1960s.
- The term 'peak oil' (or gas) refers to the year in which global production will reach its maximum level and then fall into sustained decline.
- Peak oil and gas will have social and economic impacts.

## Unit outcomes

By the end of this unit most students should be able to:

- use evidence to show that demand for oil and other sources of energy continues to rise;
- explain why the West is expected to dominate energy consumption in future;
- mark the location of the world's top oil and gas reserves on a world map;
- explain why it is difficult to increase oil production;
- define the term 'peak oil' and draw a Venn diagram to show the social and economic impacts of peak oil and gas.

## Ideas for a starter

1. Open up a discussion. Have we got enough energy? Who is using it all? Where are all the oil and gas reserves? How long will they last?
2. Use the quote from Colin Campbell on page 36 of the Student Book to introduce the idea that oil will not last forever and at some point production will decline.
3. Ask students to explain the term 'peak oil' and why it is important.

## Ideas for plenaries

1. Ask students to work in pairs to write a paragraph with the heading 'Is there enough energy?'
2. Ask students to think back over this unit and write down three questions related to what they have learned. Ask individual students to feed back.
3. Use the 'What do you think?' on page 39 of the Student Book as a plenary.
4. Add the term 'peak oil' (and gas) to the dictionary of key terms for this chapter.

# Business as usual

## The unit in brief

This is a 6-page unit which looks at our continued reliance on fossil fuels, i.e. 'business as usual'. The IEA predicts that fossil fuels will continue to dominate the energy mix for the foreseeable future – but this will come at a cost in terms of rising carbon-dioxide emissions and the threat of climate change.

The Middle East is the focus for this unit. It is a key supplier of oil and its importance is likely to increase in future – but security and stability are vital if the rest of the world wants to rely on its oil supplies. The final spread looks at the American invasion of Iraq (it is now clear to many people that the main motive was to gain access to Iraq's oil reserves) and the situation in Iran and Central Asia.

## Key ideas

- The IEA predicts that fossil fuels will continue to dominate the energy mix between 2005 and 2030.
- By 2030, the carbon-dioxide emissions associated with energy use will rise by 57%, increasing the threat of climate change.
- The Middle East is a key supplier of oil and it will become increasingly important.
- By invading Iraq, the USA may have hoped to reduce its dependence on Saudi Arabian oil and to access Iraq's oil reserves.
- Iran is a major production hotspot – an energy-producing country or region where there is political instability.

## Unit outcomes

By the end of this unit most students should be able to:

- list the IEA's predictions for energy consumption and production;
- draw a table to show the costs and benefits of continuing to rely on fossil fuels as major sources of energy;
- describe the importance of the Middle East in terms of oil supply;
- recognise that the fight for energy security can raise geopolitical tensions.

## Ideas for a starter

1. Brainstorm: What does 'business as usual' mean in the context of energy security? What are the likely costs and benefits of business as usual?
2. Show either of the photos on page 44 of the Student Book on the whiteboard, but reveal them a little at a time. Ask: Where is this soldier? Why is he there? What is the link with energy security?
3. Ask a student to read President Bush's statement at the top of page 44 of the Student Book. Ask: Where do the USA's oil imports come from? Why is this important?

## Ideas for plenaries

1. Ask students to write down as many words as they can relating to this unit and to add any key terms and their definitions to their dictionary for this chapter.
2. Use 'On your own' activity **6** on page 45 of the Student Book as the basis for a class discussion. Ask students to write up the discussion.
3. Ask students to create a mind map around the words 'Middle East'. How many ideas can they come up with in three minutes?

# 1.10 Meeting future energy needs

## The unit in brief

This is the synoptic unit for the 'Energy security' chapter. It is a 4-page unit in which students investigate some of the approaches we can adopt to meet our future energy needs and increase our energy security. It focuses on alternatives to 'business as usual' (reliance on fossil fuels) and provides students with a range of resources which look at nuclear power, renewable energy, and approaches such as conservation, recycling energy, and 'green' taxation. The unit ends with a synoptic question to give students an idea of what examiners could ask in the exam.

## Key ideas

- There are a range of approaches we can adopt to meet our future energy needs.
- There are arguments for and against nuclear power.
- The use of renewable energy is growing.
- Consumption of energy can be reduced by conservation, recycling energy, and the introduction of 'green' taxes.

## Unit outcomes

By the end of this unit most students should be able to:

- outline the ways in which we can meet our future energy needs;
- analyse the strengths and weaknesses of the pro-nuclear and pro-renewable energy lobbies;
- give examples of how the use of renewable energy is growing;
- give two examples of how energy consumption can be reduced.

## Ideas for a starter

1 Show Resource 1 from page 46 of the Student Book on the whiteboard. Ask students what the costs and benefits of nuclear power are. Should we use nuclear power as part of our future energy mix?

2 Ask students for ideas of how energy can be saved in the home. Create a spider diagram of their ideas and compare them with the diagram of the methods being used in Woking Borough Council's low-carbon home on page 48 of the Student Book.

3 Ask: What are green taxes? Would students be willing to pay more to fund projects to tackle climate change? (In 2008, a poll revealed that 7 out of 10 voters in Britain would not be willing to pay higher taxes for that purpose.)

## Ideas for plenaries

1 Students could work in pairs to plan an answer to one part of the synoptic question on page 49 of the Student Book.

2 Use the 'What do you think?' on page 49 to discuss whether it is possible for the rest of the world to follow New Zealand's lead in significantly increasing the use of renewable energy.

# 2 Water conflicts

## Chapter outline

Use this chapter outline and the introductory page of the chapter in the Student Book to give students a mental roadmap for the chapter.

## About the topic

- This topic has two underlying themes – water resources, and the conflicts between those who have water and those who do not.
- Water is a human need, but it is not evenly distributed. Physical factors help to determine supplies, whilst human factors determine how well these supplies are distributed.
- Growing demand for water does not match supply, and countries often seek supplies across national borders. The potential for conflict is therefore high.
- High demand can also lead to long-term degradation.

## About the chapter

- This chapter begins with a case study of California – looking at the issues of water supply and environmental degradation.
- It looks at the global imbalance between water supply and demand, and the global water crisis.
- It investigates the geopolitical issues related to the demand and supply of water in the Middle East.
- It deals with the links between water, wealth, and poverty, using contrasting examples from Africa and the Murray-Darling Basin in Australia.
- The price of water is the focus of the final unit – looking at the impact of China's economic growth on water supplies and water privatisation.

## Key vocabulary

There is no set list of words in the specification that students must know. However, examiners will use some or all of the following words in the examinations, and would expect students to know them and use them in their answers.

aquifer
desalination
El Niño
geopolitical
groundwater
high pressure
infiltration
irrigation
La Niña
percolation
precipitation
prevailing
privatisation
rain shadow
relief rainfall
riparian
salinity
spatial imbalance
streamflow
surface runoff
urbanisation
virtual water
water pathways
water rights
water scarcity
water stress
water wars
world water gap

The glossary at the end of this book contains many of these words and phrases. For students, the key word boxes in the chapter or the glossary at the end of the Student Book will help them with the meanings of all.

# 2.1 California calling

## The unit in brief

This 6-page unit introduces the chapter on 'Water conflicts' with a case study of the problems of supplying water to California. California is the world's sixth largest economy and for many Americans it offers a high-quality lifestyle. But the 37.7 million people who live in California are facing up to ever-increasing problems with their water supplies.

The unit takes a close look at the Colorado River Basin which supplies water to seven surrounding US states, plus Mexico, and which is under severe pressure as a water supply. Attention in California is now focusing on demand management rather than increasing overall supplies of water.

## Key ideas

- California's water supplies are threatened as a result of precipitation (lack of, and seasonal shortages), population (growth, increasing demand for water), and spatial imbalance between available water and population distribution.
- California depends on two major water supply lines (the State Water Project combined with the Central Valley Project, and the Colorado River).
- Geographical controls influence the availability of water.
- Throughout the twentieth century, agreements and treaties were needed to allocate 'fair shares' of the Colorado River Basin's water to the surrounding US states and Mexico.
- Agreements about water shares were based on flow patterns in the early twentieth century when rainfall was higher than today.
- The 2007 agreement aimed to divide the Colorado's shortages rather than to share the flow.

## Unit outcomes

By the end of this unit most students should be able to:

- understand how and why California's water supplies are threatened;
- assess the importance of the Colorado River in terms of California's water supply;
- explain how geographical controls influence the availability of water in California;
- outline the key decisions made about sharing the Colorado;
- explain why pressures are building on the Colorado;
- suggest how Californians could be encouraged to use water more sustainably.

## Ideas for a starter

1. Ask: What do you know about California? You are looking to elicit ideas about its wealth, population, climate, lifestyle, and so on. What is the link between students' ideas and the topic of 'Water conflicts'? Record ideas as a mind map on the board.
2. Show students a variety of images of California such as the photo on page 53 of the Student Book, those of Lake Powell on page 57, and others, for example, of Death Valley and urban sprawl. Ask: Why does California have a water problem?

## Ideas for plenaries

1. Use the 'What do you think?' on page 57 of the Student Book as a plenary. What would be the impact if Native Americans claimed all their water rights?
2. Use 'On your own' activity **5** on page 57 to hold a class discussion. Students can write up the discussion later in 500 words.
3. Ask students to work in pairs to write a paragraph on the problems of sharing the Colorado.

# 2.2 California – environment at risk

## The unit in brief

This 2-page unit continues the case study of California begun in Unit 2.1. It looks at the conflict between supplying water to California and safeguarding the environment. During the twentieth century, California's focus was on obtaining enough water for the State's needs without really considering the environmental consequences. The disregard for the environment, and its natural processes and habitats, is now putting all three at risk. This unit looks at the problems facing the Sacramento-San Joaquin and Colorado River Deltas, and the Salton Sea.

## Key ideas

- The environment, and its natural processes and habitats, are at risk in California as a result of the State's water acquisition.
- The Sacramento-San Joaquin River Delta faces the problems of old poorly maintained levees and river banks; endangered fish species; water-treatment works discharging chlorine into the rivers.
- The CALFED Bay-Delta Program aims to develop a sustainable long-term solution to water management and environmental problems.
- The Salton Sea is under threat from high levels of agricultural run-off; high evaporation rates; high salinity, algal blooms, and eutrophication; industrial pollution and sewage.
- The Salton Sea Restoration Project faced a choice in 2003 between maintaining the Salton Sea as an agricultural sump, or restoring its natural habitats.

## Unit outcomes

By the end of this unit most students should be able to:

- explain why the Sacramento-San Joaquin and Colorado River Deltas and the Salton Sea are in environmental trouble;
- describe the problems faced by the Sacramento-San Joaquin River Delta;
- list the economic costs and environmental benefits of the strategies being adopted by the CALFED Bay-Delta Program;
- outline the problems facing the Salton Sea;
- describe the value of the Salton Sea to an economist and to an environmentalist.

## Ideas for a starter

1. Recap on Unit 2.1: the problems of supplying water to California. Use the recap to introduce the idea that supplying water creates further problems.
2. Show students photos of the consequences of California's use of water – dried salt deposits on the shore of the Salton Sea (from page 59 of the Student Book), and dried up riverside habitat alongside the Colorado. Ask: What do these images show?

## Ideas for plenaries

1. Give students two blank maps of California. Ask them to locate and name the Sacramento-San Joaquin River Delta and the Salton Sea. They should annotate the first map with the problems faced, and the second map with solutions or choices for the future.
2. Ask students to summarise what they have learned in this unit in 35 words.

# 2.3 Global imbalance

## The unit in brief

The global imbalance between water supply and demand is the focus of these two pages. The unit investigates where our water is stored, and the issues of a world water gap, water stress, and water scarcity. It ends with some statistics on global water consumption: population growth and economic development in the twentieth century caused massive increases in water consumption, a trend which is likely to continue.

## Key ideas

- Fresh water makes up only 2.5% of the water on the Earth's surface, but only half of that is available for human consumption.
- A 'world water gap' exists between those who have, and those who do not have, sufficient clean water.
- An increase in the global population means that the number of people affected by water stress and water scarcity is expected to rise sharply.
- During the twentieth century, global water consumption increased by 600%.

## Unit outcomes

By the end of this unit most students should be able to:

- explain why so little water is available for human consumption;
- identify areas which are likely to have a water surplus or a water shortage in future;
- describe the global distribution of water scarcity;
- say why and how global water consumption is likely to increase further in the twentieth century.

## Ideas for a starter

1. Fill a large jug or bucket with 4.5 litres of water (and have a tablespoon handy). Tell students this represents all the water on Earth. Ask them how much they think represents the amount of available freshwater. (The answer is one tablespoonful.)
2. Ask individual students to read out the bullet points at the foot of page 60 of the Student Book. Use them to introduce the topic of the imbalance between water supply and demand.
3. This unit contains a wealth of diagrams and statistics – use any of them to kick off the unit.

## Ideas for plenaries

1. Ask students to explain the terms 'water stress' and 'water scarcity' (physical and economic) to their partners.
2. Use 'On your own' activity **2** on page 61 of the Student Book as a plenary.

# 2.4 Global water crisis

## The unit in brief

This 4-page unit looks at some of the reasons for the global water crisis. In 2003, the UN's World Water Development Report warned that 'the lack of freshwater is emerging as the biggest challenge of the twenty-first century'. Increasing demand and misuse is putting supplies of freshwater at risk.

The unit looks at the issues of water insecurity in different regions, sanitation, and the impacts of politics on water security, as well as the links between poverty and the lack of water quality.

## Key ideas

- The lack of freshwater is a major challenge for the twenty-first century.
- People's health, welfare, and livelihoods depend on secure supplies of freshwater, but increasing demand and misuse of water resources put these at risk.
- Improved sanitation reduces poverty and increases economic development.
- There is a strong link between poverty and a lack of access to safe water and sanitation.
- Economic development in some countries has led to a decline in water quality.

## Unit outcomes

By the end of this unit most students should be able to:

- use evidence to explain why the lack of freshwater puts people's health, welfare, and livelihoods at risk;
- draw a flowchart to show how improved sanitation reduces poverty;
- draw a simple graph to show the link between poverty and either access to safe water, or access to sanitation;
- describe, giving examples, how economic development can lead to a decline in water quality.

## Ideas for a starter

1. Show the three photos of the Aral Sea in Kazakhstan (from 1977, 1989, and 2006) at the foot of page 62 of the Student Book on the whiteboard as students enter the room. Ask: What has happened? Why? What is the likely impact of this on water supplies in the area?
2. Ask a student to read aloud Abida Bibi's report on the death of her child in the speech bubble at the foot of page 63 of the Student Book. Tell students that each year 250 000 children die in Pakistan from water-borne diseases.

## Ideas for plenaries

1. Use the 'What do you think?' on page 64 of the Student Book to generate a discussion of the issues addressed in this unit. Alternatively you could use it to generate a spider diagram of the reasons for the global water crisis.
2. Ask students to identify the three key messages they learned from this unit.

# 2.5 Water insecurity

## The unit in brief

The first two pages of this unit look at the potential for conflict over water resources around the world. It investigates how conflicts can arise between neighbouring countries and some of the key players involved in water – the WTO, TNCs, and the UN.

The second two pages of the unit focus on water stress in the Middle East and North Africa. Demand for water already exceeds natural supply in many Middle Eastern countries, with most suffering from water stress or scarcity. Turkey and Israel have become focal points as far as water management and international relations are concerned.

## Key ideas

- The risks of real water shortages are growing, and as they do, so does the potential for conflict.
- Countries need secure water supplies to further their economic development.
- In many countries control of the water infrastructure is being taken over by TNCs.
- The Middle East faces a number of issues including a scarcity of water, poor access to water, a rising youthful population, and increasing demands on water and food resources.
- Turkey and Israel have become major focal points in terms of water management and international relations.

## Unit outcomes

By the end of this unit most students should be able to:

- give examples of how one country can threaten another's water supply and create the potential for conflict;
- explain how countries can use their water supplies to further their economic development;
- explain the link between the WTO, water, TNCs, and debt relief;
- outline the issues relating to water which face the Middle East;
- give examples of how Turkey and Israel have raised tensions in the Middle East as a result of their water management.

## Ideas for a starter

1. Write the words 'Water wars' in the middle of the whiteboard. Ask for ideas about water and conflicts over water to create a spider diagram.
2. Show the graph on page 67 of the Student Book on the whiteboard – minus the caption. Ask what students think the graph is about (the last bar should give them a clue). Are they surprised that disagreements over water can lead to military acts and serious diplomatic disputes?

## Ideas for plenaries

1. Test students' understanding of this unit. Ask:
   - Why can water shortages lead to conflict?
   - Name some countries where water resources are contested.
   - How has water become big business?
   - What water issues does the Middle East face?
2. Give students five minutes to write a radio bulletin on the topic 'Water and the potential for conflict'.

# 2.6 Managing water insecurity

## The unit in brief

This is a 6-page unit which looks at how water management strategies have raised tensions in the Middle East. The first two pages focus on Turkey, and investigate the issues raised by Turkey's GAP project – the Southeastern Anatolia Project. GAP has upset Turkey's neighbours, and international funding was withdrawn as a result of some of the social and environmental issues connected with the project.

The unit includes four pages which look at Israel's water supply issues. Israel's water insecurity has led to disputes with its neighbours, but has also forced Israel to become more water-efficient and to develop strategies to manage its limited supplies.

## Key ideas

- Turkey's GAP project is an attempt to provide an integrated water and energy supply system, but it has become a geopolitical issue.
- Turkey had to amend GAP to reduce its social and environmental impacts.
- Israel's water problems revolve around a water shortage, competition for water, a growing population, and increasing and longer droughts.
- Israel's water insecurity has led to disputes with its neighbours.
- Israel's water management strategies include managing limited supplies, acquiring new supplies, and expanding virtual water supplies.

## Unit outcomes

By the end of this unit most students should be able to:

- summarise how GAP became a geopolitical issue;
- say why Turkey amended GAP to reduce its social and environmental impacts;
- list the reasons for Israel's water insecurity;
- annotate a map to show Israel's disputes with its neighbours over water;
- describe Israel's water management strategies.

## Ideas for a starter

1. Show the photo of Hasankeyf in Turkey at the top of page 71 in the Student Book on the whiteboard. Ask students to imagine this is their town. Tell them that it is going to be flooded and 34 000 people will lose their homes as part of the GAP project. What is their reaction? How do they feel?
2. Put the Turkish President's speech about water rights in the speech bubble on page 70 of the Student Book on the whiteboard. Do students agree with the President's view that Turkey can do whatever it likes?

## Ideas for plenaries

1. With books closed, ask students to define geopolitics and virtual water. Ask for one example of a geopolitical issue and one example of the trade in virtual water.
2. Use the 'What do you think?' on page 75 of the Student Book as a plenary.
3. Use 'Over to you' activity **2** on page 75 to start a debate about shared water resources. Ask students to sum up the debate in writing in 300 words.

# 2.7 Water, wealth and poverty – 1

## The unit in brief

This is the synoptic unit for the 'Water conflicts' chapter. It is a 6-page unit in which students begin to investigate the link between water, poverty, and wealth. It focuses on the future of Africa's water resources and provides students with a range of resources to enable them to assess the impact of the water crisis on Africa. The unit investigates competition for water and includes resources which look at three different approaches to coping with the challenges Africa faces: the trade in virtual water, major irrigation schemes, and small-scale projects. The unit ends with a synoptic question to give students an idea of what examiners could ask in the exam.

## Key ideas

- Africa is facing a water crisis.
- There is a growing demand, and competition, for Africa's water resources.
- Africa's trade in virtual water creates water shortages and a range of other problems.
- Major irrigation schemes, such as Toshka, are controversial and have disadvantages.
- Small-scale projects, such as those supported by WaterAid, enable communities to achieve a better quality of life and escape from poverty.

## Unit outcomes

By the end of this unit most students should be able to:

- assess how serious the issue of water stress is in Africa as a whole, and the Nile Basin in particular;
- explain why African countries see mega-projects as a key part of their economic development programmes;
- assess the likely environmental and socio-economic impacts of the proposals for coping with water shortages in African countries.

## Ideas for a starter

1. Show students one or more of the cash crops grown in African countries for export, for example, green beans, mange tout, baby sweetcorn, flowers, or coffee. Ask: What are the likely impacts of growing these on Africa's water supply?
2. Ask individual students to read out the speech bubbles and text boxes on page 78 of the Student Book. Use this to start a discussion on the use of water in African countries.
3. Show the bottom map from page 76 of how rainfall is predicted to change by 2030 (minus the caption) on the whiteboard. Ask students what they think the map shows. What impact will this have?

## Ideas for plenaries

1. Ask students to write the phrase 'Water, wealth, and poverty' in the middle of the page and create a mind map around it. How many ideas can students come up with in two minutes?
2. Students could work in pairs to plan an answer to one part of the synoptic question on page 81 of the Student Book.

# 2.8 Water, wealth and poverty – 2

## The unit in brief

This 6-page unit looks at the water problems facing the Murray-Darling Basin in Australia. The MDB is an enormous asset in terms of its agricultural production, but is under threat from increasing and competing demands for water.

The MDB has experienced environmental degradation as a result of changes in agriculture and piecemeal approaches to regulating the river system. Conflicts arise where stakeholders have different views about how water in the MDB should be managed and used. In 2007 the Australian government announced the National Plan for Water Security focusing on the MDB in an attempt to secure its future.

## Key ideas

- The MDB is a huge asset but is under threat from increasing and competing demands for water.
- The MDB is experiencing environmental degradation as a result of changes in agriculture and a piecemeal approach to regulating the river system.
- The MDB has many stakeholders who have different views about the way the resource is managed, which can lead to conflict.
- One of the biggest changes in the MDB was a policy called 'The Cap' which created environmental flows.
- The National Plan for Water Security was implemented in 2008 to provide an integrated management system to restore the MDB.

## Unit outcomes

By the end of this unit most students should be able to:

- explain why and how the MDB is under threat;
- describe the environmental degradation occurring in the MDB;
- complete a conflict matrix to evaluate the values and attitudes of the main stakeholders in the management of the MDB;
- list the benefits of environmental flows;
- outline the main aspects of the National Plan for Water Security.

## Ideas for a starter

1. Ask: What can you tell me about the MDB? Where is it? What problems does it face? Why is water such a big issue there? (If you used *AS Geography for Edexcel*, students may have already studied the MDB in the context of topic 'Extreme weather' and so be familiar with the area.)
2. Read out the BBC news extract on page 158 of the *AS Geography for Edexcel* Student Book. Lead on to a discussion of water in Australia and the need for management.

## Ideas for plenaries

1. The 'Over to you' activity **2** on page 87 of the Student Book, completing the conflict matrix and assuming the role of one of the MDB stakeholders, could be used as a plenary.
2. The 'What do you think?' on page 87 could be used as a discussion to draw the unit to a close. Ask: How do management decisions made today differ from those made in the past?
3. Ask students to tell their partners the two key things they learned from this unit and then another two things that are interesting but less important.

# 2.9 The price of water

## The unit in brief

The first two pages of this unit investigate China. They look at how China's economic growth has increased its demand for water, the impacts that this is having, and how technology – by way of the Three Gorges Dam and the South-to-North Water Transfer Project – is being used to try to solve the country's water supply problems and secure its economic prosperity.

The final two pages look at the issues raised by privatising water supplies, factors affecting future demand and supply of water, and how countries might respond to future challenges.

## Key ideas

- China's economic growth has drastically increased its need for water.
- China's economic growth is threatening environmental well-being.
- China is using technological fixes to try to solve the country's water supply problems.
- Increasing water privatisation has resulted in increasing costs, or a lack of connection to supplies, for some people.
- How countries respond to future challenges of demand and supply of water may reflect one of three scenarios: business-as-usual; technology, economics, and privatisation; values and lifestyle.

## Unit outcomes

By the end of this unit most students should be able to:

- describe how China's economic growth has increased its need for water;
- decide whether it is possible to achieve economic development without sacrificing the environment;
- assess the strengths and weaknesses of the South-to-North Water Transfer Project;
- explain why water privatisation has created problems;
- comment on how the challenges posed by the three future water scenarios might be overcome.

## Ideas for a starter

1. Show the photo on page 88 of the Student Book on the whiteboard. Tell students this 'shows' the Yangtze in Chongqing, China. It is the longest river in Asia and the third longest river in the world. Ask: What has happened, and why?
2. The price of water. Ask students to imagine that they live in Cochabamba, Bolivia. The year is 1999 and the city's water supply has just been privatised. As a result, water prices have risen. They are poor and face a choice between paying 20% of their wages on water supplies, or feeding their children. What will they do? (Read page 90 of the Student Book to find out what happened in this real-life example.)
3. Brainstorm: the price of water. What ideas does this phrase conjure up for students?

## Ideas for plenaries

1. Use the 'What do you think?' on page 91 of the Student Book as a plenary.
2. Ask students to write 'Water conflicts' in the middle of the page and create a mind map around it. How many ideas can students come up with in two minutes?
3. What price water? People can live for about two months without food, but less than a week without water. Should it ever be privatised? Debate the issue.

# 3 Biodiversity under threat

## Chapter outline

Use this chapter outline and the introductory page of the chapter in the Student Book to give students a mental roadmap for the chapter.

**3.1** **The great Alaskan wilderness** The Arctic National Wildlife Refuge in Alaska and the threat facing its biodiversity and people's well-being

**3.2** **Global distribution of biodiversity** What influences the variations in, and distribution of, biodiversity

**3.3** **Biodiversity under threat** Investigating some of the threats to biodiversity

**3.4** **The Daintree rainforest** Focusing on the Daintree rainforest, and looking at the tropical rainforest ecosystem and ecosystem services

**3.5** **Threats to the Daintree** Threats to the Daintree rainforest ecosystem and nutrient cycles

**3.6** **Can the threats to biodiversity be successfully managed?** Investigating different ways of managing the threats to the Daintree rainforest

**3.7** **Mangroves** What mangroves are, what mangrove ecosystems are like, and mangroves and ecosystem services

**3.8** **Threats to mangroves** Some of the threats facing mangroves and ecosystem processes

**3.9** **Managing the threats to mangroves** Exploring a range of management Options to halt the loss of biodiversity in mangroves

**3.10** **Biodiversity – what is the future?** Assessing the future of biodiversity (Synoptic unit)

## About the topic

- There are two main themes to this topic; biodiversity as a key resource, providing valuable goods and services to people, and its importance to human and ecological wellbeing.
- It is about the pressures placed on biodiversity by economic development.
- People value biodiversity differently. When valued and exploited for its economic potential, rather than for its ecological value, biodiversity is threatened.
- Reconciling economic development and the need to maintain biodiversity is a challenge, together with the threats posed by climate change and alien species.

## About the chapter

- This chapter tackles the topic of biodiversity under threat through a case study approach, backed up with theory where needed.
- It begins with a case study of the Arctic National Wildlife Refuge in Alaska, and investigates the threats it faces from the possibility of oil development.
- It investigates the global distribution of biodiversity, threats to biodiversity, and biodiversity hotspots.
- Case studies of the Daintree rainforest and mangroves look at ecosystems under threat, the services they provide, and how the threats can be managed.
- The chapter concludes by investigating the future of biodiversity.

## Key vocabulary

There is no set list of words in the specification that students must know. However, examiners will use some or all of the following words in the examinations, and would expect students to know them and use them in their answers.

alien species
anaerobic
aquaculture
biodiversity
biodiversity hotspots
biomass
biome
buy-back
carbon sequestration
conservation
coral bleaching
co-region
cultural services (in terms of biodiversity)
ecosystem diversity
endemic
endemism
eutrophication
genetic diversity
Gersmehl's nutrient cycles
global orchestration
inter-tidal area
leaching
lenticels
mangrove
net primary productivity
organic productivity
permafrost
photosynthesis
pivotal areas
pneumatophores
prop roots
provisioning services (in terms of biodiversity)
Ramsar sites
red, black and white/grey mangroves
regulating services (in terms of biodiversity)
species diversity
supporting services (in terms of biodiversity)
sustainable use
sustainable yield
trophic
zonation

The glossary at the end of this book contains many of these words and phrases. For students, the key word boxes in the chapter or the glossary at the end of the Student Book will help them with the meanings of all.

# 3.1 The great Alaskan wilderness

## The unit in brief

This 6-page unit uses a case study of the Arctic National Wildlife Refuge (ANWR) in Alaska to introduce the chapter on 'Biodiversity under threat'. The ANWR is the most biologically diverse Arctic region in the world, but it is believed that oil lies below its Coastal Plain.

The unit looks at the players interested in whether the Coastal Plain should be developed for oil or not, and the arguments for and against development. The future of the Coastal Plain may come down to politics and whether or not the USA can do without its oil.

## Key ideas

- Biodiversity means biological diversity. It is the variety of all forms of life on Earth.
- The Coastal Plain is a fragile ecosystem where oil exploration and development has not yet been allowed.
- There are a wide range of people and organisations interested in whether the Coastal Plain should be developed or not.
- There are arguments for and against developing the Coastal Plain.
- Action has been taken to protect biodiversity and the Coastal Plain.
- If there is a future oil shortage, the USA may need to reconsider development of the Coastal Plain.

## Unit outcomes

By the end of this unit most students should be able to:

- define biodiversity;
- complete a conflict matrix to show different groups which are likely to agree or disagree over the development of the ANWR;
- complete a table to assess the impacts of developing the ANWR;
- say what action has been taken to protect biodiversity and the Coastal Plain;
- explain why the USA may decide to develop the Coastal Plain in future.

## Ideas for a starter

1. Show photos of a range of wildlife which lives on the Coastal Plain of the ANWR. Reveal them one at a time. Photos could include caribou (see page 94 of the Student Book), polar bears, snow geese (see page 98), tundra swans, grizzly bears, wolves, etc. Ask: Where do all these creatures live? Why is this place important?
2. Ask: What does biodiversity mean? How many species are there worldwide? Why are we concerned about biodiversity?
3. Show the photo of the Coastal Plain from page 96 of the Student Book on the whiteboard. How would students describe the image? Does this look like a special place worth saving from development?

## Ideas for plenaries

1. Divide students into two groups. One group should take on the role of the Inupiat people of North Alaska, and the other group the Gwich'in. The two groups should argue for or against development of the Coastal Plain for oil.
2. Use the 'What do you think?' on page 98 of the Student Book to debate whether it would be worth going ahead with the development of the Coastal Plain.
3. You could use 'Over to you' activity **2** on page 99 of the Student Book as a plenary.
4. Get students to begin a dictionary of key terms for this chapter. Begin with 'biodiversity' and 'organic productivity'.

# 3.2 Global distribution of biodiversity

## The unit in brief

This 3-page unit looks at the variation in, and distribution of, biodiversity. Biodiversity is greatest in the tropics and declines towards the poles. Tropical rainforests contain over 50% of the world's species in just over 7% of the world's land area. They are also the world's most productive biome.

The unit includes a Background box which looks at the factors and processes influencing biodiversity. These include climate (precipitation, temperature, light intensity, and winds), endemism, and human activity.

## Key ideas

- There are global variations in biodiversity. Biodiversity is greatest in the tropics and declines towards the poles.
- Biomes are large global ecosystems which vary in terms of their biodiversity.
- Biodiversity hotspots are areas with high concentrations of biodiversity.
- Pivotal areas are those with concentrations of hotspots.
- Tropical rainforests are the most productive biome.
- There are a range of factors and processes which influence biodiversity including climate, endemism, and human activity.

## Unit outcomes

By the end of this unit most students should be able to:

- explain the global variations in biodiversity in terms of climate, endemism, and human activity;
- understand that biomes are not uniform in terms of their biodiversity;
- define the terms 'biodiversity hotspots' and 'pivotal areas';
- explain the importance of biodiversity hotspots.

## Ideas for a starter

1. Ask a student to read out the newspaper extract about the giant rat on page 100 of the Student Book – but minus the last paragraph. Why is this area described as the 'lost world'?
2. Show the map on page 100 on the whiteboard. What does this show? What factors or processes influence biodiversity or species richness?

## Ideas for plenaries

1. With books closed, ask:
   - What are biodiversity hotspots?
   - Give me an example of a pivotal area.
   - What is net primary productivity?
   - What factors and processes influence biodiversity?
2. Get students to add the key terms for this unit to their dictionary of key terms.
3. Use 'Over to you' activity **2** on page 102 of the Student Book as a plenary.

# 3.3 Biodiversity under threat

## The unit in brief

In this 3-page unit students investigate some of the threats to biodiversity. The main drivers, or threats, are seen as habitat change, over-exploitation, the introduction of alien species, pollution, and climate change. It is thought that by the end of the twenty-first century, climate change and its impacts may be the dominant driver of biodiversity loss.

The unit looks at three biodiversity hotspots: the Caucasus, Southwest Australia, and the Atlantic Forest, and the range of threats they face.

## Key ideas

- Extinction is a natural event, but the rate of extinction is accelerating as a result of human activity.
- The main threats to biodiversity are habitat change, climate change, the introduction of alien species, over-exploitation, and pollution.
- Climate change and its impacts may be the dominant drivers of biodiversity loss by the end of the twenty-first century.
- Biodiversity hotspots are under threat from human activity.

## Unit outcomes

By the end of this unit most students should be able to:

- understand that extinction is a natural event, but that the rate of extinction is accelerating as a result of human activity;
- give examples of how habitat change, climate change, the introduction of alien species, over-exploitation, and pollution threaten biodiversity;
- describe the current trends in the threats to biodiversity;
- describe the distribution of threatened hotspots and the threats that they face.

## Ideas for a starter

1. Show an image of a dodo (see page 103 of the Student Book), a Calvaria major tree, a cassowary, and a map of Mauritius on the whiteboard. Ask students which is the odd one out, and why. (The answer is the cassowary, but students need to see if they can work out the link between the other three. The paragraph at the top of page 103 will provide you with the answer.)
2. Ask a student to read out the paragraph on habitat change on page 103 of the Student Book, and time them. It will probably take about 30 seconds. Then tell students that in those 30 seconds approximately 50 acres of rainforest were destroyed – and the destruction continues. Ask students what they think the impacts of rainforest destruction will be on biodiversity.

## Ideas for plenaries

1. How far do students agree with the statement that 'By the end of this century, climate change and its impacts may be the dominant driver of biodiversity loss?'
2. Ask students to sum up what they have learned in this unit in 35 words.
3. 'The main threats to biodiversity are…'. Go round the class, adding information without hesitation or repetition.

# 3.4 The Daintree rainforest

## The unit in brief

This 4-page unit begins the case study of the Daintree rainforest. The Daintree rainforest is in northern Queensland, Australia and is the oldest rainforest in the world. Its biodiversity is influenced by climate and endemism.

The unit includes a Background box which explains ecosystem services – the benefits that people get from ecosystems. They can be classified as: provisioning services, regulating services, cultural services, and supporting services. It then looks at the benefits which the Daintree, like all rainforest ecosystems, provides.

## Key ideas

- The Daintree rainforest in northern Queensland is part of the Wet Tropics, and has the highest levels of biodiversity and regional endemism in Australia.
- The Daintree rainforest is the oldest rainforest in the world.
- Climate and endemism have both influenced biodiversity in the Daintree rainforest.
- Ecosystem services are benefits that people get from ecosystems.
- The Daintree rainforest provides a range of benefits and services.

## Unit outcomes

By the end of this unit most students should be able to:

- use evidence to explain why the Daintree is an important ecosystem;
- decide whether climate or endemism have been more important in influencing biodiversity in the Daintree;
- classify ways in which ecosystems can benefit people as provisioning, regulating, cultural and supporting services;
- give examples of the benefits and services that the Daintree rainforest provides.

## Ideas for a starter

1. Ask students if anyone has been to Australia and the Daintree rainforest. Ask them to describe the rainforest to the rest of the class. What makes the ecosystem so special?
2. As an alternative to starter **1**, darken the room. Ask students to close their eyes and imagine they are in a tropical rainforest. What sights and sounds can they see and hear? What makes this ecosystem such a fragile environment?
3. Ask: What is a World Heritage Site? Name some examples. What criteria would you use to decide what makes a World Heritage Site? How do students' ideas compare with the list of criteria on page 106 of the Student Book?

## Ideas for plenaries

1. Ask: What are ecosystem services? How can they be classified? Give me some examples of the services that the Daintree rainforest provides.
2. Ask students to write the 'Daintree rainforest' in the middle of the paper, and create a mind map around the words. How many ideas can students come up with in two minutes?
3. 'Over to you' activity **2** on page 109 of the Student Book could be used as a plenary. Students will need to work in groups to create spider diagrams of how different people perceive the Daintree and then complete a conflict matrix. This will help them to start thinking about the threats to the Daintree which Unit 3.5 looks at.

# 3.5 Threats to the Daintree

## The unit in brief

In this 5-page unit students find out about threats to the Daintree rainforest ecosystem. The biggest threat facing the Daintree is tourism. Tourism is the world's fastest growing industry and, by 2004, visitor numbers to the Daintree had increased to 500 000. Port Douglas is close to the Daintree and has been affected by the increasing numbers of tourists to the region. There are fears that development will spread to the Daintree itself.

The unit includes two Background boxes – one on deforestation (one of the causes of habitat change, and a threat to biodiversity) and one on the recycling of nutrients (one of the basic processes that ecosystems depend on).

## Key ideas

- The biggest threat facing the Daintree rainforest is tourism and the number of tourists is growing rapidly.
- Port Douglas has developed as tourist numbers have increased.
- There is concern that development will spread into the Daintree.
- Deforestation is a major threat to biodiversity.
- Ecosystems depend on the recycling of nutrients.

## Unit outcomes

By the end of this unit most students should be able to:

- draw a spider diagram to show how tourism and development pose threats to the Daintree;
- explain what the biggest threats to the Daintree are;
- suggest how the ferry, the lack of mains electricity, and local services can limit development in the Daintree;
- classify the causes of deforestation as social, environmental, and economic;
- draw a diagram to show how deforestation affects the nutrient cycle.

## Ideas for a starter

1. Recap on Unit 3.4: Where is the Daintree? Why is it special? What benefits and services does it provide?
2. Go to www.pddt.com.au/daintree-national-park.php and show the Port Douglas and Daintree screen on the interactive whiteboard. Access the website to find out what tourists can do in the Daintree.
3. Brainstorm: How can tourism and development threaten biodiversity? Create a spider diagram of the ideas that students come up with.

## Ideas for plenaries

1. If you used starter **3**, return to the spider diagram. Ask students to complete 'Over to you' activity **1** on page 114 of the Student Book and compare their ideas now with the ideas they came up with at the start of the unit.
2. 'Tourism is good for the local economy. It creates jobs and is worth millions of dollars.' Divide the class into two groups – one in favour of tourism, and one which wants to limit tourism. Debate the idea.
3. Ask students to write down as many ideas as they can related to this unit. Ask them to explain the most important three ideas to their partners.

# 3.6 Can the threats to biodiversity be successfully managed?

## The unit in brief

This 3-page unit concludes the case study of the Daintree rainforest. It investigates the range of strategies and players involved in managing biodiversity in the Daintree. It looks at the roles played by the Wet Tropics Management Authority, Douglas Shire Council, the Rainforest Co-operative Research Council, and a number of other players.

The unit encourages students to consider the conflicts that can arise between different players as a result of their differing economic and environmental interests.

## Key ideas

- There is a range of strategies and players involved in managing biodiversity in the Daintree, but sometimes they come into conflict.
- The Wet Tropics Management Authority is responsible for managing the Wet Tropics as a World Heritage Site.
- In 2008, the Daintree became part of Cairns Regional Council.
- The Rainforest Co-operative Research Council proposed a range of strategies to protect the environment and build a sustainable community.
- Other players involved in preserving the Daintree include the Australian Rainforest Foundation, the Wildlife Preservation Society of Queensland, and the Australian Tropical Research Foundation.

## Unit outcomes

By the end of this unit most students should be able to:

- complete a conflict matrix to show how far the organisations which have an interest in the Daintree agree or disagree;
- assess how far the conflicts are a case of economic versus environmental interests;
- explain why the abolition of Douglas Shire Council in 2008 could increase development in the Daintree;
- describe the range of strategies proposed by the Rainforest Co-operative Research Council to protect the environment and build a sustainable community;
- give examples of the projects that other players are involved in.

## Ideas for a starter

1. Show the photo of Cairns from page 115 of the Student Book (or any other photos which show development in Cairns) on the whiteboard. Tourism is the biggest threat facing the Daintree. Ask students how they think the threats to biodiversity in the Daintree can be managed. Record their responses on a spider diagram.
2. Ask students what they think an 'indicator species' is. How can indicator species be protected?

## Ideas for plenaries

1. Use the 'What do you think?' on page 115 of the Student Book as a plenary.
2. How could the Daintree change now that it has become the responsibility of Cairns Regional Council?
3. Divide the class into four groups. Each group should represent one of the organisations which have an interest in the Daintree: the Wet Tropics Management Authority, Douglas Shire Council, the Rainforest Co-operative Research Council, and the Australian Rainforest Foundation. In a discussion, each group should put forward their views on managing biodiversity. How far do they agree?

# 3.7 Mangroves

## The unit in brief

This is a 4-page unit which begins a case study of mangrove ecosystems. The unit begins with a graphic account of the value mangroves have in acting as a coastal barrier against tsunami, cyclones, and storm surges. It continues by looking at what mangroves are, where they are located, the vegetation, and their productivity.

The unit concludes by looking at the goods and services that mangroves provide, and at the economic value of mangroves.

## Key ideas

- Intact mangroves act as barriers to tsunami, cyclones, and storm surges.
- Mangroves are essential to marine, freshwater, and terrestrial biodiversity.
- Mangroves are located along the tropical and sub-tropical coasts of Africa, Australia, Asia, and the Americas.
- Mangroves are highly productive habitats.
- Mangroves provide a range of goods and services, and have an economic value.

## Unit outcomes

By the end of this unit most students should be able to:

- recognise the value of mangroves in terms of coastal protection;
- explain why mangroves are essential to marine, freshwater, and terrestrial biodiversity;
- describe the distribution of mangroves;
- decide which of the ecosystem services that mangroves provide are of value to different people;
- explain why the economic value of cleared mangroves declines.

## Ideas for a starter

1. Show the photo at the top of page 118 of the Student Book on the whiteboard, and read out the text on destroying Asia's coastal barrier to the class to highlight the importance of mangroves.
2. Name this ecosystem. Tell students it includes fauna such as manatees, crab-eating monkeys, fishing cats, monitor lizards, Royal Bengal tigers, and mud skipper fish. It is full of sticky mud and biting insects – and it smells. What is it, and where do you find it?

## Ideas for plenaries

1. Ask students to tell their partners what an eco-region is and what aquaculture is, and then add these to their dictionary of key terms for this chapter.
2. Use the 'What do you think?' on page 118 of the Student Book to start a debate.
3. You could use 'Over to you' activity **1** on page 121 of the Student Book as a plenary. Students should decide which ecosystem goods and services are valuable to different people and why their attitudes to the value of biodiversity might vary.

# 3.8 Threats to mangroves

## The unit in brief

This 5-page unit continues the case study of mangrove ecosystems. It looks at the rate of mangrove loss and the threats that mangroves face. One of the major threats is the rise in aquaculture – largely due to shrimp farming (over 50% of the mangrove loss in Asia has been due to aquaculture). Mangroves face other threats, though, and these include climate change, over-harvesting, over-exploitation, tourism, oil exploration and development, and environmental degradation.

The unit includes two Background boxes which look at the importance of mangroves in terms of carbon sequestration, and the process of energy flows in ecosystems.

## Key ideas

- Over 50% of the world's original mangrove forests have been lost and this has a major impact on biodiversity.
- In Asia, over 50% of mangrove loss has been due to increasing aquaculture.
- Aquaculture has some benefits, but many of the impacts have been negative.
- Mangroves face threats from climate change, over-harvesting, over-exploitation of marine life, tourism, oil exploration and development, and environmental degradation.
- Mangrove loss means a loss of carbon sequestration.
- Ecosystems depend on the flow of energy.

## Unit outcomes

By the end of this unit most students should be able to:

- draw a diagram to show the links between mangrove destruction and economic activity;
- list the positive and negative impacts of shrimp farming;
- draw a diagram to show the main threats to mangroves;
- explain the link between mangrove destruction and carbon sequestration;
- assess how mangrove destruction affects the entire ecosystem.

## Ideas for a starter

1. Find a photo which shows a landscape of rice paddies – preferably in Thailand. Show it on the whiteboard side by side with the photo on page 122 of the Student Book which shows a landscape covered in shrimp ponds. Tell students that the photos show the landscape before and after the expansion of aquaculture. Ask them what they think the impacts of aquaculture are likely to be.
2. Show the table of Asia's shrimp production from page 123 of the Student Book on the whiteboard. Ask: What do these figures show? Why has shrimp production increased so much?
3. Brainstorm: What are likely to be the main threats to mangroves? Why? Record students' ideas as a spider diagram.

## Ideas for plenaries

1. Ask students to re-read the text about Dulah Kwankha at the top of page 123 of the Student Book. Ask them to put themselves in his shoes. He may have more money to spend as a result of shrimp farming, but has it improved his quality of life?
2. Use the 'What do you think?' on page 126 of the Student Book as a plenary.
3. Ask students to sum up what they have learned in this unit in 40 words.

# 3.9 Managing the threats to mangroves

## The unit in brief

This is a 3-page unit which concludes the case study on mangrove ecosystems. There are many players who are trying to halt the loss of mangroves, and to protect and restore them, and this unit explores the following: community development organisations such as Yadfon; regional organisations such as TVE Asia Pacific; NGOs like Wetlands International; EU-funded initiatives, such as the Coastal Biodiversity in Ranong project; and intergovernmental agreements like Ramsar Sites.

## Key ideas

- There are a range of management Options to halt the loss of biodiversity in mangroves.
- Management Options operate at a range of scales from community development organisations to intergovernmental agreements.
- The different approaches towards achieving sustainability have advantages and disadvantages.
- The different approaches vary in terms of their costs, benefits, and effectiveness.

## Unit outcomes

By the end of this unit most students should be able to:

- summarise the range of management Options available to halt the loss of biodiversity in mangroves;
- be aware that management Options operate at a range of scales;
- complete a table to show the advantages and disadvantages of the approaches towards achieving sustainability of mangroves;
- rank the projects in terms of their costs, benefits, and effectiveness.

## Ideas for a starter

1. Recap: What are mangrove ecosystems like? Why are they important? What are the threats that mangroves face?
2. Show the photo of the dugong on page 127 of the Student Book on the whiteboard. Ask: What is this strange-looking creature? What is its relevance?
3. Ask students, in pairs, to come up with as many words as they can to do with the threats facing mangroves.

## Ideas for plenaries

1. Ask students to unpick this statement: 'Human well-being and ecological well-being are inter-linked.'
2. Students could work in groups of three to complete 'On you own' activity **3** on page 129 of the Student Book, with each student completing one map. They could copy and share their maps with others in the group.
3. Play 'Just a minute'. The topic is 'Managing the threats to mangroves'. Students take it in turns to talk on the topic without hesitation or repetition.

# 3.10 Biodiversity – what is the future?

## The unit in brief

This is the synoptic unit for the chapter on 'Biodiversity under threat'. It is a 4-page unit in which students explore a range of resources in order to assess the future of biodiversity. These include resources on the Millennium Development Goals, the Convention on Biological Diversity, the Millennium Ecosystem Assessment and the four possible scenarios put forward for the future by the MEA. The unit ends with a synoptic question to give students an idea of what examiners could ask in the exam.

## Key ideas

- The sustainable management of biodiversity is critical to human well-being and is central to achieving the Millennium Development Goals.
- The Convention on Biological Diversity was committed to achieving a significant reduction of biodiversity loss by 2010.
- Short-term goals and targets such as the Convention on Biological Diversity may not be sufficient for the conservation and sustainable use of biodiversity and ecosystems.
- The MEA has suggested a range of Options to tackle biodiversity loss and environmental degradation.
- The MEA has put forward four possible future scenarios which focus on alternative approaches to sustaining ecosystem services.

## Unit outcomes

By the end of this unit most students should be able to:

- understand that the sustainable management of biodiversity is critical to human well-being;
- assess the importance of biodiversity in the Millennium Development Goals;
- explain why long-terms goals and targets are needed for the sustainable management of biodiversity;
- describe the Options suggested by the MEA to tackle biodiversity loss and environmental degradation;
- analyse the four MEA scenarios and assess which is most likely to be achieved.

## Ideas for a starter

1. Write this phrase on the whiteboard: 'Poverty and biodiversity are intimately linked'. Create a spider diagram around the phrase. Ask students why they are linked, and how.
2. Use any of the quotes on the four pages in the Student Book to kick off this unit. They convey powerful messages.
3. Ask: What are the Millennium Development Goals? Why are they relevant to biodiversity?

## Ideas for plenaries

1. Students could work in pairs to plan an answer to one part of the synoptic question on page 133 of the Student Book.
2. Use the 'What do you think?' on page 132 of the Student Book as a plenary.

# 4 Superpower geographies

## Chapter outline

Use this chapter outline and the introductory page of the chapter in the Student Book to give students a mental roadmap for the chapter.

**4.1 Changing world order** How the balance between superpowers can change over time

**4.2 Who are the superpowers?** What does 'superpower' mean? How to decide whether a country is a superpower or not

**4.3 The earliest superpowers** How the British Empire became the world's first major superpower in modern times, and theories behind colonialism

**4.4 Cold War and superpower rivalries** The development of the USA and USSR as twentieth-century superpowers, and the role of international financial organisations

**4.5 Colonialism – gone but not forgotten?** Has gaining independence from British colonial rule given Ghana true freedom?

**4.6 China – economic superpower?** How China is gaining influence as a new economic superpower

**4.7 Russia – the re-emerging superpower?** The collapse of the USSR and the re-emergence of Russia as a global player

**4.8 Superpower influences over nation states** How superpowers play a key role in international decision-making

**4.9 Cultural superpowers** How superpower influence extends to 'global culture' (Synoptic unit)

## About the topic

- This topic has two themes: power, and the way that it develops over time in some countries and regions, and changing balances of power in the current world.
- It is about how some nations and players have a disproportionate influence over regional and global decision-making.
- The geography of power develops over time; some nations gain power and influence, while others lose it.
- The nature of power varies, from direct to indirect control, e.g. through trade, culture, and resources.
- In future, the BRICs (Brazil, Russia, India and China) are likely to become increasingly powerful.

Teachers and their students will also find it useful to cross-refer to Chapter 5 'Bridging the development gap' to help understand further the impacts of some of the world's superpowers.

## About the chapter

- This chapter investigates what 'superpower' means and what factors we can use to assess whether a country is a superpower or not.
- It looks at how the balance between superpowers can change over time – how the British Empire became the world's first superpower in modern times, and how the USA and USSR developed into twentieth-century superpowers.
- Theories behind colonialism are covered, along with a case study of Ghana to investigate whether independence from colonial rule has given Ghana true freedom.
- Case studies of China and Russia provide examples of a new economic superpower and the re-emergence of a global player.
- The chapter ends by looking at cultural superpowers – how superpower influence extends to 'global culture'.

## Key vocabulary

There is no set list of words in the specification that students must know. However, examiners will use some or all of the following words in the examinations, and would expect students to know them and use them in their answers.

capitalism
Cold War
colonialism
command government
commodity
commodity trading exchanges
communism
cultural imperialism
dependency theory
devaluation
development theory
direct influence
disparity
electronic colonialism
EU enlargement
evangelism
futures market
*glasnost*
hegemony
ideology
International Monetary Fund (IMF)
Mackinder's heartland theory
market economy
modernisation theory
modernism
NATO
neo-colonialism
*perestroika*
primary products
privatisation
secondary products
social Darwinism
superpower
tariff
tariff escalation
USSR
value-added
vicious cycle of development (or poverty)
virtuous cycle of development
World Bank
World Trade Organisation (WTO)

The glossary at the end of this book contains many of these words and phrases. For students, the key word boxes in the chapter or the glossary at the end of the Student Book will help them with the meanings of all.

# 4.1 Changing world order

## The unit in brief

This 2-page unit introduces the chapter on 'Superpower geographies'. It begins with the fall of the Berlin Wall in 1989 and the collapse of the communist bloc – events which students may not associate initially with Geography, but which demonstrate the concept of changing world order. These were important events in the context of superpowers and the last two decades have seen enormous political and economic change. We now face an era of new superpowers with the economic rise of the EU and rises in the economies of the BRICs – Brazil, Russia, India, and China.

## Key ideas

- The balance between superpowers can change over time.
- Between 1989 and 1991, the communist bloc based on the USSR was dismantled.
- The past two decades have brought enormous political and economic change.
- A new era of superpowers is beginning.

## Unit outcomes

By the end of this unit most students should be able to:

- understand how the balance between superpowers can change over time;
- draw a spider diagram to show the relevance of the changes which occurred between 1989 and 1991;
- describe some of the political and economic changes of the past two decades;
- draw a table to show some of the impacts of the new superpowers in terms of raw material consumption.

## Ideas for a starter

1. Show the video clip from the BBC News website on the fall of the Berlin Wall, which occurred on 9 November 1989. To find this, type 'On This Day – 1989: Fall of the Berlin Wall' into the search facility on BBC's news website www.bbc.co.uk/news.
2. Brainstorm: 'What has the fall of the Berlin Wall to do with Geography, and why?' Ask particularly for any feedback from students who study Politics or History.
3. Show the two photos from pages 136 and 137 of the Student Book on the whiteboard. Ask students what the connection is between them. (The answer is superpowers. One photo represents the fall of the Berlin Wall, the dismantling of the communist bloc and the decline of the USSR (the world's second superpower after the USA); the other represents the rise of the new superpowers.)
4. What do superpowers have to do with the study of Geography? Ask students to discuss or complete a spider diagram and feed back to the rest of the class.

## Ideas for plenaries

1. Use 'Over to you' activity **1** on page 137 of the Student Book as a plenary.
2. Get students to work in pairs to write a paragraph on how the balance of superpowers can change over time.
3. Use the 'What do you think?' on page 136 of the Student Book as a plenary. This will get students thinking about what 'superpower' actually means. This will be good preparation for Unit 4.2.

# 4.2 Who are the superpowers?

## The unit in brief

In this 4-page unit students investigate what the word 'superpower' means, and the criteria which can be used to assess whether or not a country can be seen as a superpower. Put simply, a superpower is a country with power and influence over others. Only the USA remains as a major world power now, but that situation is rapidly changing. There are a number of factors which help a country to gain power and influence, and the unit looks at the following: physical size and geographical position; resources; population size; economic power and influence; military force; dominant belief systems.

## Key ideas

- A superpower is a country with the capacity to project power and influence over others.
- Physical size and position determine the area over which a country has potential influence.
- Resources can be critical to economic development and therefore to the emergence of superpowers, but it is only when countries withhold them that they become a means of influence.
- A large population can act as a spur to economic growth, but it is not crucial.
- The world's top 12 economies influence much of what happens in the global economy.
- Military force has historically been a major influence in determining political power.
- Dominant belief systems include religion and global capitalism, and are a critical feature of superpower influence.

## Unit outcomes

By the end of this unit most students should be able to:

- define the term 'superpower';
- understand what makes a country a superpower;
- rank the criteria which are most important in creating the world's superpowers;
- justify other criteria for determining a superpower.

## Ideas for a starter

1 Brainstorm: What is a superpower? What criteria would you use to decide? Who do you consider are the world's current superpowers, and why?

2 You could use any of the maps in this unit to start it off, but especially those from the worldmapper website (see the top of page 139 and foot of page 140). The two in this unit show countries in proportion to the volume of iron-ore exports, and the size of their military forces. The world looks very different when viewed in this way.

## Ideas for plenaries

1 Use the 'What do you think?' on page 141 of the Student Book as a plenary. If superpowers are essential for keeping world peace, which ones do we mean?

2 What further information might help to determine who the world's superpowers are a) now, and b) likely to be in 20 years' time?

# 4.3 The earliest superpowers

## The unit in brief

In this 4-page unit students learn how the British Empire became the world's first major superpower in modern times. They look at how colonialism took place between the fifteenth and nineteenth centuries, as European countries sought land overseas to expand their political control, particularly in Africa and the Americas.

Colonialism arose from an ideology or set of beliefs, but the desire to colonise and dominate the world was also driven by philosophies such as modernism, Mackinder's heartland theory, and evangelical Christianity and social Darwinism.

## Key ideas

- At the beginning of the twentieth century, the British Empire was the biggest global superpower.
- Colonialism took place between the fifteenth and nineteenth centuries as European countries sought land overseas to expand their political control.
- Maintaining colonial rule depended on military power.
- Colonialism arose from an ideology or set of beliefs.
- The desire to colonise and dominate the world was also driven by philosophies, such as modernism, Mackinder's heartland theory, and evangelical Christianity and social Darwinism.

## Unit outcomes

By the end of this unit most students should be able to:

- explain the term 'colonialism';
- explain why Britain believed it was right to obtain colonies;
- describe how colonialism developed between the fifteenth and nineteenth centuries;
- prepare a presentation for or against the philosophies of modernism, Mackinder's heartland theory, or evangelical Christianity, and social Darwinism.

## Ideas for a starter

1 Play a recording of 'Rule Britannia' as students enter the room. Ask: Why is this music playing? What does this have to do with the topic of superpowers, and the previous lesson?

2 Show the map on page 142 of the Student Book (or any map which shows the extent of the British Empire) on the whiteboard. Ask: What are the areas in pink? What does it mean in reality for a country to colonise another? Why did Britain want to control large areas of the world?

## Ideas for plenaries

1 Ask students to work in groups of three to write a paragraph in their own words to explain one of: modernism, Mackinder's heartland theory, and evangelical Christianity and social Darwinism.

2 Ask: How can colonialism be defined?

3 Use the 'What do you think?' on page 143 of the Student Book as a plenary.

# 4.4 Cold War and superpower rivalries

## The unit in brief

This is a 6-page unit which begins by looking at the development of the USA and USSR as twentieth-century superpowers. By 1900, the USA had taken over from Britain as the world's largest economy and, by the 1960s, the USSR had become the world's second economic superpower. The unit explores how the USSR and USA have influenced world geography, and includes Background boxes on capitalism and communism, and modernisation theory in relation to the USA. It ends by exploring the mechanisms by which superpowers try to assert their global influence, through the development and workings of the International Monetary Fund and the World Bank, with examples of their roles in the recovery and rebuilding of Singapore and Japan.

## Key ideas

- Both the USA and USSR possess resources which have influenced economic development.
- By 1900, the USA was the world's largest economy and superpower.
- By the 1960s, the USSR had become the world's second economic superpower, as a result of a series of five-year plans which increased economic production.
- During the Cold War, the USA, believing in capitalism, was opposed to the Soviet communist government. Similarly, the USSR believed in a global domination of communism.
- During the 1970s, the USSR influenced Africa (and other areas of the world) through the provision of military assistance, financial aid, and attempts to destabilise countries.
- The USA aimed to prevent communism by increasing economic development based upon modernisation theory. The establishment of the International Monetary Fund and the World Bank helped the USA to achieve their aim.

## Unit outcomes

By the end of this unit most students should be able to:

- compare the USA and USSR in terms of how they became superpowers;
- compare the strengths and weaknesses of capitalism and communism as theories;
- describe how the USSR influenced countries in Africa in the 1970s;
- summarise how the USA aimed to prevent communism by increasing economic development based on modernisation theory;
- explain the roles of the International Monetary Fund and the World Bank.

## Ideas for a starter

1. Show a clip from a film or BBC series such as 'Tinker, Tailor, Soldier, Spy'.
2. Ask: What can you tell me about the Cold War? What was the Iron Curtain?
3. Find a news article in which one of the world's major economies, such as China, is buying resources or assisting in the development or affairs of another country. Discuss: What is happening here? Why might it be happening? Why might one country want to influence the affairs of another?

## Ideas for plenaries

1. Three ideas for discussion: use 'Over to you' activity **2** on page 151 of the Student Book and discuss the strengths and weaknesses of communism and capitalism as theories as a class; use 'Over to you' activity **4** on page 151; use the 'What do you think?' on page 150.
2. Ask: Explain communism and capitalism. Give me an example of a communist country. How do most countries operate now?
3. Create a mind map around the phrase 'Cold War and superpower rivalries'.

# 4.5 Colonialism – gone but not forgotten?

## The unit in brief

This 6-page unit is a case study of Ghana in which students learn whether gaining independence from British colonial rule has given Ghana true freedom. Although Ghana gained independence from Britain in 1957, it still depends on decisions made by the world's wealthy countries – a situation known as neo-colonialism. The unit examines the effects of neo-colonialism on Ghana by examining the cocoa trade. Ghana has no control over the price of cocoa – commodity traders, overseas tariffs, and the WTO all influence the price. Ghanaian farmers have now begun to form co-operatives, such as Kuapa Kokoo, to take control over prices and improve people's lives.

## Key ideas

- Ghana gained independence from Britain in 1957, but still depends on decisions made by wealthy nations – a system referred to as 'neo-colonialism'.
- The cocoa trade has always been important to Ghana but the price of cocoa is decided by commodity traders, overseas tariffs and the WTO, rather than by Ghana.
- Dependency theory argues that the cause of poverty in developing countries is their reliance on developed economies.
- Development theory argues that the world's richer nations are wealthy because of their control over commodities and raw materials imported from developing countries. This in turn prevents developing countries from accumulating capital to invest in manufacturing.
- The WTO exists in theory to promote free and fair trade between the world's nations; in fact, many of its benefits accrue to developed rather than developing countries.
- Ghanaian farmers have now begun to form co-operatives to take control over prices and improve people's lives. This could represent a more sustainable model of economic development.

## Unit outcomes

By the end of this unit most students should be able to:

- define the term 'neo-colonialism';
- explain how far commodity markets, overseas tariffs, and the WTO work for Ghana's benefit;
- explain dependency theory and development theory, and how far organisations such as the WTO exist to promote these ideas;
- explain how communities in Ghana can benefit from co-operatives such as Kuapa Kokoo.

## Ideas for a starter

1 Show the table comparing Ghana's indicators of development on page 152 of the Student Book on the whiteboard. Ask pairs of students to discuss: a) What progress has Ghana made in the 20 years between 1987 and 2007? b) What progress has not been made?

## Ideas for plenaries

1 Get students to re-read the speech bubble on page 154 of the Student Book, giving the views of William Korampong, a Ghanaian cocoa farmer, and look at the photo of the woman on page 157. Ask: Where, how, and why are decisions being made that affect these people's future and livelihoods?

2 Ask students to work in pairs to make notes in preparation for the report they need to write for 'On your own' activity **6** on page 157 of the Student Book.

3 Ask students to write down as many words as they can to do with the effects of neo-colonialism on Ghana.

# 4.6 China – economic superpower?

## The unit in brief

In this 6-page unit students learn how China is gaining influence as a new economic superpower. The unit begins by looking at the town of Karratha in Western Australia, which is experiencing a boom. The reason for the boom is the local iron ore, most of which is sold to China to fuel its economic expansion. China's demand for raw materials is so great that it is buying companies to secure its future supplies and prices of raw materials and fuel. But China's economic growth has been achieved at a high environmental and social cost, and the unit also looks at these issues.

## Key ideas

- Since the early 1980s, China's economy has doubled in size every eight years, and China has the largest sustained GDP growth in history.
- China's demand for raw materials to feed its growing economy is soaring.
- China is trying to guarantee supplies of raw materials such as iron ore by buying the companies that supply them. Similarly, China is trying to guarantee its fuel supplies.
- China's economic growth has been achieved at high environmental and social cost.
- China has limited conventional military reach, though it is estimated that it is the world's third largest defence spender.

## Unit outcomes

By the end of this unit most students should be able to:

- give examples of China's increasing demand for raw materials;
- explain how China is attempting to guarantee raw materials and fuel supplies;
- draw a spider diagram to show what future threats could occur to China's continued economic growth;
- assess whether economic development is possible without causing environmental damage;
- give examples of the social costs of China's economic growth.

## Ideas for a starter

1. Show students images of the Pilbara and its iron ore. Ask: How is this place connected to other countries? Students could also explore in the atlas what other resources Australia possesses, and who might be interested in these in the current global economic climate, and why.
2. Read out, or put on the whiteboard, the list of bullet points on page 158 of the Student Book, and those in the Background box on page 161. Ask: How has China grown so fast?
3. Write the phrase 'China is going shopping' on the board. Brainstorm ideas on what students think China's rapidly growing economy needs. Record ideas as a spider diagram. Economics students could develop this further – they will know that what China is doing is an example of achieving a vertical monopoly, i.e. monopolising the supply chain throughout the production line, in order to guarantee supply and influence price levels.

## Ideas for plenaries

1. Ask students to re-read the text on page 162 of the Student Book and associated newspaper reports on human rights issues. Then use the 'What do you think?' on page 163 as a plenary.
2. Ask students to work in pairs to develop an argument on 'China – economic superpower?'

# 4.7 Russia – the re-emerging superpower?

## The unit in brief

This 4-page unit looks at the collapse of the USSR and the re-emergence of its largest republic, Russia, as a global player. In 1987, the USSR's communist leader Mikhail Gorbachev introduced a number of changes to combat political and social discontent and economic stagnation. These changes led ultimately to the collapse of the USSR in 1991.

During the 1990s, Russia's state-run economy collapsed with massive economic, social, and environmental impacts. Since 1999, Russia's economy has been recovering, helped largely by its oil and natural gas reserves, but wealth remains in the hands of a very few individuals.

## Key ideas

- In the 1980s, the USSR was one of two global superpowers, with a command-style government.
- Changes introduced by Mikhail Gorbachev in the 1980s, known as 'glasnost' and 'perestroika', led ultimately to the collapse of the USSR in 1991.
- The collapse of Russia's economy in the 1990s had huge economic, social, and environmental impacts.
- Russia's large oil and natural gas reserves, controlled largely by a few very wealthy individuals, have helped to fund its economic recovery since 1999.
- Russia has one of the world's largest military forces, though its prosperity has had little impact on defence.

## Unit outcomes

By the end of this unit most students should be able to:

- describe how the policies of *perestroika* and *glasnost* led to the collapse of the USSR in 1991;
- draw spider diagrams to show the economic, social, and environmental impacts of changes in Russia since 1991;
- list the factors that have helped Russia to re-emerge as a global power;
- decide whether Russia now deserves the title 'superpower'.

## Ideas for a starter

1. Show students a photo of KGB agent Alexander Litvinenko, who died after being poisoned with Polonium-210. The UK suspected a fellow former KGB agent. Ask: What does this tell us about control in Russia a) now, and b) under the KGB?
2. Find some text (try the BBC website) on what life behind the iron curtain was like, for example, in East Germany. Ask students to assess the pluses and minuses of communism for people who lived under it. This could then lead to a discussion of which two things (i.e. economic freedom and social freedom) might have been perceived by some as necessary to improve the life or economic production of the former communist countries.
3. Show the graph of the Russian economy on page 167 of the Student Book on the whiteboard to start the unit. Ask: How are Russia's fortune's changing? Does its growing economy mean it can re-emerge as a superpower?

## Ideas for plenaries

1. Use 'Over to you' activity **3** on page 167 of the Student Book as a plenary.
2. Get students to have another look at the text and related information on Russia's military influence. Then use the 'What do you think?' on page 166 of the Student Book as a plenary.

# 4.8 Superpower influences over nation states

## The unit in brief

In this 4-page unit students learn how superpowers play a key role in international decision-making and the organisations that they establish in order to assist this process. On the first two pages, they investigate the expanding influence of the EU and the factors that will affect its future influence, as well as NATO and its changing role.

The final two pages look at Russia and its influence over former regions of the USSR, such as Ukraine (and the gas disputes in 2006–7 and 2007–8), and the South Caucasus.

## Key ideas

- Europe is growing more unified and powerful, both politically and economically.
- The EU has expanded dramatically since it was set up in 1957, but there are a number of factors which are affecting, and will continue to affect, its future influence.
- NATO has grown and changed since it was set up in 1949.
- The absorption of former communist countries in the EU and NATO has increased tensions with Russia.
- Three factors help to explain the conflict over gas between Ukraine and Russia.
- The South Caucasus is geographically significant both because of its location and the discovery of natural gas in the Shah Deniz gas field.

## Unit outcomes

By the end of this unit most students should be able to:

- list the benefits of joining both the EU and NATO, as perceived by those countries wishing to join;
- list the factors which will affect the future influence of the EU;
- suggest what NATO and the EU have to gain and lose by admitting more countries;
- define the terms 'direct' and 'indirect influence';
- summarise the factors that help to explain the conflict over gas between Ukraine and Russia;
- explain the geographical significance of the South Caucasus.

## Ideas for a starter

1. Show a video clip of the Eurovision Song Contest as students enter the room. Ask students why they think this is being shown. Eurovision is more about a country's identity and aspiration – the desire to 'be' European – than a desire to compete in a bizarre musical contest.
2. Brainstorm: What is Europe? How do we define it? How has Europe's influence expanded?

## Ideas for plenaries

1. Get students to work in pairs to make notes on how membership of inter-governmental organisations gives some countries political and economic power.
2. Use the 'What do you think?' on page 170 of the Student Book as a plenary.
3. Ask students to select and investigate in pairs which countries are the most likely to join a) the EU, and b) NATO, and why. This could be developed as a role play where students present a case for their country, and a panel asks questions and then makes decisions about who should and should not be admitted.

# 4.9 Cultural superpowers

## The unit in brief

This is the synoptic unit for the 'Superpower geographies' chapter. It is a 4-page unit in which students investigate how superpower influence extends to 'global culture'. It asks whether the world is becoming Americanised, and provides students with a wide range of resources on the world's biggest brands, the film industry, TV viewing, the music industry, and News Corporation, to assess the impacts of cultural Americanisation, and whether it is beneficial. It asks to what extent other countries (such as India or the UK) also have a claim to be 'global cultural superpowers'. The unit also includes a Background box on cultural imperialism.

It ends with a synoptic question to give students an idea of the styles of questions that examiners could ask in the exam.

## Key ideas

- Cultural imperialism refers to either an enforced spread of its culture by a larger power, or the voluntary adoption of a foreign culture by other people.
- The dominant global culture now is American, which spreads by broadcasting its TV, films, and music throughout the world; other countries such as India can also lay a claim.
- The impacts of cultural Americanisation may not be beneficial for everyone.
- Cultural influence lies in the hands of a few very large and wealthy companies such as News Corporation.
- Cultural influences from elsewhere can either enrich or threaten the local culture.
- The expanding global economy has led to a global culture, spread by information technology.

## Unit outcomes

By the end of this unit most students should be able to:
- understand how cultural influence can spread;
- define cultural imperialism;
- assess the extent to which cultural imperialism can be beneficial;
- assess how healthy it is for the production of most music, television, and film to be in the hands of so few.

## Ideas for a starter

1. There is a plethora of films and music you could use to start this unit off. Consult the list of top-ten largest-grossing films on page 172 of the Student Book and the tables on culture and the music industry on page 173. Show clips of the films or play tracks from some of the albums as students enter the room. Ask students to think critically about why you are showing the film/playing the music.
2. Show any of the tables from pages 172 and 173, or the worldmapper map on page 174 of the Student Book to kick off this unit. Ask questions about the data shown.
3. Show the photo on page 174 of the Student Book on the whiteboard. Where do students think this is likely to be? What does it show us?

## Ideas for plenaries

1. Ask: What is cultural imperialism? What is a hegemony?
2. Get students to re-read the text on the News Corporation on page 175 of the Student Book. Then use the 'What do you think?' on page 175 as a plenary.
3. Students could work in pairs to plan an answer to one part of the synoptic question on page 175.

# 5 Bridging the development gap

## Chapter outline

Use this chapter outline and the introductory page of the chapter in the Student Book to give students a mental roadmap for the chapter.

**5.1 Food in crisis** How the global food crisis affects some people and countries more than others

**5.2 Identifying the development gap** How indicators are used to highlight the development gap

**5.3 Living on the wrong side of the gap** What the development gap means for those affected by it, using Uganda as an example

**5.4 How we see the world** How people's attitudes towards global development and other countries are linked to the development gap

**5.5 South Africa – the widening gap** How the development gap can be found within one country, and results from political systems dating back to colonialism

**5.6 Booming Bangalore** How rapid development in one Indian city has exposed cultural differences between people

**5.7 Ways forward 1: the importance of trade** How important trade is in enabling economic development to take place

**5.8 Ways forward 2: aid or investment?** Investigating whether aid or investment is the best way of funding development (Synoptic unit)

**5.9 Ways forward 3: the Millennium Development Goals** What the Millennium Development Goals are and how far progress has been made in achieving them

## About the topic

- This topic has two themes; it is about the development gap and its origins, and how it might be reduced.
- It offers theoretical perspectives about development and how it can best take place.
- It is about ways in which the wealth of western economies contrasts with the continuing poverty of some peoples and nations in the developing world.
- The 'gap' is generally increasing.

Teachers and their students will also find it useful to cross-refer to Chapter 4 'Superpower geographies' to help understand further how and why the development gap arose.

## About the chapter

- This chapter investigates what the development gap is and how indicators can be used to highlight it.
- It uses the example of Uganda to illustrate what the development gap means for those affected by it.
- A case study of South Africa illustrates how the development gap can be found within one country, whilst Bangalore is included to show how rapid development can expose cultural differences between people.
- The theories of 'core and periphery' (a geographical model) and 'economic man' (a classical economic theory) are included to show how European and, later, American thinking came to dominate the world.
- The chapter ends by looking at the variety of ways in which economic development can take place and the development gap closed – through trade, aid or investment, and the Millennium Development Goals.

## Key vocabulary

There is no set list of words in the specification that students must know. However, examiners will use some or all of the following words in the examinations, and would expect students to know them and use them in their answers.

aid
altruism
apartheid
bilateral aid
bottom-up development
capital-intensive
commodities
core and periphery
debt service
development
development gap
disparity
economic man
Euro-centric
food security
formal economy
Gini Co-efficient
Gross Domestic Product (GDP)
Gross National Income (GNI)
Human Development Index (HDI)
informal economy
investment
labour-intensive
Millennium Development Goals (MDGs)
multilateral aid
multiplier effect
neo-liberalism
Non-Governmental Organisation (NGO)
OPEC
out-sourcing
Peace Index
per capita
privatisation
Purchasing Power Parity (PPP)
reciprocity
segregation
Structural Adjustment Packages (SAPs)
tied aid
top-down development
trade liberalisation

The glossary at the end of this book contains many of these words and phrases. For students, the key word boxes in the chapter or the glossary at the end of the Student Book will help them with the meanings of all.

# 5.1 Food in crisis

## The unit in brief

This 2-page unit introduces the chapter on 'Bridging the development gap' by looking at how the global food crisis affects some people and countries more than others. The dramatic rise in food prices from 2006 onwards and ensuing food crisis illustrates both the development gap between rich and poor worldwide, and the lack of food security in some developing countries. The unit concludes by looking at the Philippines which used to be self-sufficient in rice, but which by 2008 was the world's biggest importer.

## Key ideas

- Food prices rose sharply after 2006 as a period of economic growth and low inflation in the world's developed countries stalled.
- The food crisis illustrates the development gap – the social and economic disparity between the wealthy and poor.
- Developing countries have poorer food security than developed countries.
- Rapidly rising food prices have devastating effects on vulnerable families in developing countries.
- Until the 1990s, the Philippines was self-sufficient in rice, but by 2008 it was the world's biggest importer.

## Unit outcomes

By the end of this unit most students should be able to:

- classify the causes of rising grain prices as physical and human;
- explain how the food crisis illustrates the development gap;
- define food security and draw a spider diagram to show how the causes of rising grain prices make food security worse;
- explain why the Philippines is no longer self-sufficient in rice.

## Ideas for a starter

1. Show the text extracts about bananas, wine, and oil prices from page 178 of the Student Book on the whiteboard. Ask students what the connection is between these items and development.
2. Brainstorm: What is the development gap? What is food security? Give some examples that illustrate the development gap and food security.
3. Show the photo of rice being sold under army supervision in Manila from the foot of page 179 of the Student Book on the whiteboard. Tell students where the photo is and that it is not a refugee camp. Ask why they think selling rice needs army protection.

## Ideas for plenaries

1. Get students to compile a dictionary of key terms for this chapter. Begin with 'the development gap' and 'food security'.
2. Use the 'What do you think?' on page 179 of the Student Book as a plenary.
3. Ask students to spend five minutes writing a paragraph on the connection between the development gap and food security. They should write it in their own words.

# 5.2 Identifying the development gap

## The unit in brief

This is a 4-page unit in which students learn how indicators are used to highlight the development gap. It begins by looking at the Brandt Report from 1981, which showed the North-South divide. It explains how the development gap is measured using economic indicators, such as GDP and GNI, and human development indicators such as the HDI developed by the UN. It explains that Brandt's North-South divide now seems simplistic as rapid development since 1981 has made the world more complex. But one thing hasn't changed and that is that Africa is being left further behind.

## Key ideas

- The Brandt Report published in 1981 identified the development gap between the wealthy 'North' and poorer 'South'.
- Two indicators are commonly used to measure economic development – GDP and GNI.
- Human development is measured using the HDI.
- Rapid development since 1981 has made Brandt's North-South divide seem simplistic.
- The UN now uses four levels of income to classify the world's nations.
- Growth since 1981 has enabled some countries to develop rapidly but Africa is being left behind.

## Unit outcomes

By the end of this unit most students should be able to:

- recognise that rapid development since 1981 means that Brandt's original North-South divide is now outdated;
- draw a table to show the advantages and disadvantages of using GDP, GNI, and HDI as indicators of development;
- use maps of GDP and HDI to identify those countries which have changed since 1981, and those which have not;
- say how the UN now measures income to classify nations.

## Ideas for a starter

1. Use any of the maps on pages 180–83 of the Student Book, either on their own or in combination to start off this unit.
2. Recap: What is the development gap? How do we measure development? What indicators can we use?
3. Show the photo on page 180 of the Student Book on the whiteboard. Ask: What type of economy does this show? (The answer is that the street traders are part of an informal economy.) Figures from this type of economy are not included in development measures, but why not?

## Ideas for plenaries

1. Ask students what the terms 'development', 'GDP', 'GNI', 'PPP', and 'HDI' mean. Get students to add them to their dictionary of key terms for this chapter.
2. Ask students to identify three key ideas they have learned from this unit.
3. Give students five minutes to write a paragraph on identifying the development gap.

# 5.3 Living on the wrong side of the gap

## The unit in brief

This 6-page unit uses Uganda as a case study. It investigates what the development gap means for those affected by it. It uses the example of a newborn baby to look at life expectation and education in Uganda on the first two pages. The next two pages investigate the debt crisis and how Uganda got into debt, along with the Highly Indebted Poor Countries Initiative – an attempt to reduce debt for the poorest countries. The final two pages look at the positive impacts that debt cancellation has had on Uganda and the significance of educating girls.

## Key ideas

- For the majority in Uganda life expectancy is low, infant mortality rates are high, and living conditions are poor.
- Primary education is free; although secondary education is expensive, investment in secondary education for girls can have huge social impacts.
- Uganda's debt burden dates back to the 1970s and 1980s.
- Most of Uganda's debt was cancelled through the HIPC debt relief initiative in 2000.
- Structural Adjustment Packages imposed by the IMF meant Uganda had to make cuts in heath and education budgets in the late 1980s and early 1990s.
- The cancellation of most of Uganda's debt has had major impacts.

## Unit outcomes

By the end of this unit most students should be able to:

- draw two spider diagrams to show facts about life expectation and health, and education in Uganda;
- draw a flow chart to show how Uganda got into debt, how the money became available to borrow, and how debt mounted;
- summarise how the HIPC initiative has helped to reduce debt in Uganda and other countries;
- explain the impact SAPs had on the poor in Uganda in the late 1980s and early 1990s;
- describe the benefits of debt cancellation for Uganda.

## Ideas for a starter

1. Show the photo of the newborn baby on page 184 of the Student Book on the whiteboard. This baby was born in 2007 in northern Uganda. Ask students how they think this baby's life will differ from theirs. They should think about health, life expectation, education, and so on.
2. Brainstorm: How did many developing countries get into debt? What do students know about the HIPC initiative?
3. Show the photo of Bob Geldof from page 188 of the Student Book on the whiteboard, and ask a student to read out the speech bubble above the photo. Ask: What was Live 8? What was the connection between Live 8, the G8, and debt cancellation?

## Ideas for plenaries

1. Use the 'What do you think?' on page 189 of the Student Book to start a discussion about global debt.
2. Get students to add the terms 'debt service', 'Structural Adjustment Packages', and 'Highly Indebted Poor Countries Initiative' to their dictionary of key terms for this chapter.

# 5.4 How we see the world

## The unit in brief

In this 4-page unit students learn about how people's attitudes towards global development and other countries are linked to the development gap. The first two pages look at how we see the world by focusing on map projections – using the Mercator and Peters projections as classic examples. It is not only through maps that European, and later, American thinking came to dominate the world. The final two pages look at two theories which explain how they also became economically dominant – through a geographical model known as 'core and periphery', and through a classical economic theory known as 'economic man'.

## Key ideas

- Mercator has been the dominant map projection for over 400 years and has resulted in people seeing the world as dominated by Europe and North America.
- The Peters projection distorts the shape of places, but shows the correct areas.
- Two theories explain how European and American thinking came to dominate the world:
  - core and periphery (developed by Wallerstein): core regions drive the world economy, peripheral areas lie at the other extreme;
  - economic man (developed by Adam Smith): assumes that people act in ways that benefit themselves.

## Unit outcomes

By the end of this unit most students should be able to:

- list the advantages and disadvantages of the Mercator and Peters projections in terms of accuracy, usefulness, and appearance;
- explain the core and periphery model, giving examples;
- outline the advantages and disadvantages of Smith's theory of 'economic man'.

## Ideas for a starter

1. Ask: How can we produce a flat map from a sphere? Students may have done this before, but it is worth repeating. Divide the class into small groups and give each group an orange – to represent Earth. They need to peel the orange (in one go) and try the lay the peel flat – as if making a map projection. What happens?
2. Show the two map projections on pages 190 and 191 of the Student Book on the whiteboard. Ask: What are these projections? How long have they been around? How do they show the world?
3. Show the image of the world from space on page 191 of the Student Book on the whiteboard, along with other images which show different parts of the Earth at the centre (try Google images 'world from space' to find other images to use). Ask students how these images alter the way we see the world.

## Ideas for plenaries

1. Play 'Just a minute'. The topic is 'How we see the world'. Students take it in turns to talk on the topic for up to a minute without hesitation or repetition.
2. Students work in pairs. One should write a paragraph explaining the 'core and periphery' model, the other should write a paragraph explaining the theory of 'economic man'.
3. Use the 'What do you think?' on page 193 of the Student Book as a plenary.

# 5.5 South Africa – the widening gap

## The unit in brief

This is a 4-page unit in which students explore how the development gap can be found within one country – South Africa. The unit includes a Background box which tells students about South Africa under apartheid – from the days of colonialism up to free elections in 1994. Despite the ANC's success in those elections, economic changes since 1994 have made inequalities in South Africa even worse. Inequality shows itself through regional inequality, ethnic inequalities, and crime and personal safety.

## Key ideas

- South Africa has a development gap which results from political systems dating back to colonialism.
- After the 1994 elections, the new ANC government wanted to attract overseas investment for the expansion of mining.
- Since 1994, South Africa's economic growth has created few jobs.
- The impacts of economic changes since 1994 have made inequalities in South Africa even worse.
- One of the ANC government's successes has been the provision of housing and clean water.
- South Africa's inequality shows itself through regional inequality, ethnic inequalities, and crime and personal safety.

## Unit outcomes

By the end of this unit most students should be able to:

- use evidence to explain how South Africa's development gap dates back to colonialism;
- explain why South Africa's growth has created few jobs;
- list the reasons why inequalities have become worse since 1994;
- explain how inequality shows itself in South Africa;
- identify five priorities to improve life for both black South Africans, and all South Africans.

## Ideas for a starter

1. Brainstorm: apartheid. Do students understand what the word means? What do students know about apartheid in South Africa?
2. Show the two photos from page 196 of the Student Book on the whiteboard. Ask students where they think these photos are. How do they illustrate the development gap?
3. Show the photo of people protesting from page 194 of the Student Book. Tell students that these people are protesting about a lack of jobs and housing and are objecting to refugees. Ask: What has this got to do with geography and the development gap?

## Ideas for plenaries

1. With books closed ask: What do the terms 'capital intensive', 'labour intensive', and 'apartheid' mean? Ask students to add them to their dictionary of key terms for this chapter.
2. You could use 'Over to you' activity **2** on page 197 of the Student Book as a plenary.
3. Ask students to write down as many words to do with the development gap in South Africa as they can.

# 5.6 Booming Bangalore

## The unit in brief

This 4-page unit focuses on Bangalore, where traditional values sit alongside explosive urban growth. Bangalore is India's centre of new technology and has grown rapidly. It has the highest average incomes in India and jobs are plentiful, but India's caste system still exists, and there are major disparities between rich and poor. The unit includes a Background box which explains India's caste system. Bangalore's growth poses challenges for the future in terms of housing, transport, and energy. Bangalore's city government is planning to decentralise and to disperse IT jobs elsewhere in India.

## Key ideas

- Bangalore is India's centre of new technology, banking, finance, and the knowledge economy.
- Bangalore has the highest average incomes in India and jobs are plentiful.
- India's caste system (a religious and social class system) persists, and disparities exist between rich and poor.
- Despite changes and positive discrimination India's caste system is still widespread.
- Bangalore faces a number of future challenges.

## Unit outcomes

By the end of this unit most students should be able to:

- explain how and why Bangalore has grown so rapidly;
- list the social, economic, and environmental benefits and problems brought by Bangalore's growth;
- understand how India's caste system works and the changes to it;
- decide how Bangalore should meet its future challenges.

## Ideas for a starter

1. Show the photo of traffic congestion in Bangalore from page 198 of the Student Book on the whiteboard. With books closed, ask students where the place is and to locate it on a blank map.
2. Ask a student to read out the article on 'Bangalore's night soil collectors' from page 199 of the Student Book. Ask: What are Dalits? What can you tell me about India's caste system?
3. Has anyone in the class been to Bangalore? If so, get them to describe it to the rest of the class. What does it look like? How does it feel? What does it sound like?

## Ideas for plenaries

1. Use the 'What do you think?' on page 201 of the Student Book to discuss out-sourcing. Who gains? Who loses from out-sourcing?
2. Use 'On your own' activity **6** on page 201 to hold a class debate on the caste system and discrimination.
3. Ask students to sum up what they have learned from this unit in 35 words.

# 5.7 Ways forward 1: the importance of trade

## The unit in brief

This is a 4-page unit which picks up on the story of Uganda begun in Unit 5.3. It investigates trade in Uganda from its independence from Britain in 1962 to the situation today. In 1962 Uganda's economy was strong, but by 1987 it was devastated and in serious debt. The IMF agreed to help, but Uganda underwent a programme of structural adjustment that involved trade liberalisation which had major impacts on the country.

The unit looks at the WTO, Fair Trade (which is now helping some farmers), and Economic Partnership Agreements with the EU. Uganda still faces problems before it can trade equally with the EU.

## Key ideas

- In the long term, trade is far more effective than aid in enabling countries to develop.
- Uganda is trying to shift its economy from one based on primary exports to one with growing secondary and tertiary sectors.
- The IMF lent money to Uganda on condition that it underwent a programme of structural adjustment, which has had some major impacts.
- Uganda joined the WTO in 1995.
- Increasing numbers of farmers grow coffee for the Fair Trade market.
- In 2001, Economic Partnership Agreements began with the aim of promoting sustainable development in developing countries, but there are problems.

## Unit outcomes

By the end of this unit most students should be able to:

- explain why trade is far more effective than aid in enabling countries to develop;
- describe the advantages and disadvantages for Uganda of encouraging manufacturing, trade liberalisation, EPA's, government support for industry, and Fair Trade;
- draw a spider diagram to show the negative impacts of structural adjustment on Uganda;
- explain why joining the WTO has not benefited Uganda;
- outline the problems Uganda faces under the Economic Partnership Agreements.

## Ideas for a starter

1. Recap on Unit 5.3. Ask: What have you already learned about Uganda in terms of the development gap?
2. Ask: What are commodities? What is structural adjustment? What is trade liberalisation?
3. Show the photo of Difasi Namisi from page 204 of the Student Book on the whiteboard. What do students think he has in the basket? What are the benefits of growing Fair Trade goods?

## Ideas for plenaries

1. Get students to work in pairs to create notes in preparation for answering the exam question on page 205 of the Student Book.
2. Use the 'What do you think?' on page 205 as a plenary.
3. Ask students to add the key terms for this unit to their dictionary of key terms for this chapter.

# 5.8 Ways forward 2: aid or investment?

## The unit in brief

This is the synoptic unit for 'Bridging the development gap'. It is an 8-page unit in which students investigate whether aid or investment is the best way of funding development. It begins by looking at different types of development – top-down and bottom-up – and how each type of development is funded through aid or investment, or a combination of both. It provides students with a range of resources which are examples of investment and/or aid projects. The unit ends with a synoptic question to give students an idea of what examiners could ask in the exam, and which asks them to evaluate the projects in this unit.

## Key ideas

- Development projects aimed at closing the development gap can be top-down or bottom-up.
- Development projects, such as the Pergau Dam, Malaysia; NGO projects in Barlonyo, Uganda; and the Kingship project, Moldova, can be funded by aid.
- Development projects, such as water provision in Ghana, can be funded by investment.
- Development projects, such as the Akosombo Dam and Ghana's aluminium smelter, can be funded by a combination of aid and investment.
- Some political policies, such as land reform in Zimbabwe, are designed to redistribute wealth and close the development gap.

## Unit outcomes

By the end of this unit most students should be able to:

- understand the difference between top-down and bottom-up projects;
- explain how development projects are financed;
- understand that some policies are adopted for their popularity rather than their effectiveness;
- outline the criteria used to evaluate development projects;
- evaluate a range of development projects;
- recommend how future development projects should be managed.

## Ideas for a starter

1. Recap on Unit 5.7. Ask: What is the importance of trade in terms of closing the development gap?
2. Brainstorm to create a spider diagram on aid. What can students remember from GCSE courses about aid and investment in development projects?
3. Show students the photo of people obtaining water from a dirty pool on page 209 of the Student Book. Tell them that lack of access to safe water is a major problem in some developing countries and can keep people poor. How many ideas can they come up with for improving access to clean water?

## Ideas for plenaries

1. Ask students to write the phrase 'Ways forward: aid or investment' in the middle of the page and create a mind map around the phrase. How many ideas can students come up with in two minutes?
2. Students could work together to plan an answer to the first part of the synoptic question on page 213 of the Student Book.

# 5.9 Ways forward 3: the Millennium Development Goals

## The unit in brief

This final 4-page unit looks at the Millennium Development Goals (MDGs). The eight MDGs were agreed in 2000 to provide a set of development targets for the world to reach by 2015. Every UN member signed up, making it the largest ever multinational attempt to rid the world of extreme poverty. The unit lists the MDGs and their targets, and includes two case studies – of Bangladesh and Uganda – in order to assess how far progress has been made in achieving the goals.

## Key ideas

- Eight MDGs were agreed in 2000 to provide a set of development targets for the world to reach by 2015.
- There have been success stories in achieving the MDGs, but current improvements are not sufficient to meet MDG targets by 2015.
- The 'Call to Action' was launched by UN Secretary General Ban Ki-moon and Prime Minister Gordon Brown to hasten progress.
- Bangladesh has made progress towards achieving the MDGs, but challenges still remain.
- Uganda has made good progress so far on some MDGs, but others are way off target.

## Unit outcomes

By the end of this unit most students should be able to:

- draw a web diagram to show how the MDGs are linked;
- describe the challenges faced by Bangladesh in meeting the MDGs;
- summarise Uganda's progress towards achieving the MDGs;
- assess how likely it seems that the MDGs will be met by 2015.

## Ideas for a starter

1. Ask individual students to read out the African children's hopes for the new millennium on page 214 of the Student Book. What other things might children have hoped for at the start of the new millennium?
2. Brainstorm: What are the MDGs? Name some of them. How will they help to close the development gap?
3. Ask students to give you 10 facts about Bangladesh and 10 facts about Uganda.

## Ideas for plenaries

1. Use the 'What do you think?' on page 216 of the Student Book as a plenary.
2. Ask students to identify five key things they have learned from this chapter.
3. Ask students to work in pairs. Give them ten minutes to create a news item on how the MDGs can help to bridge the development gap.

# 6 The technological fix?

## Chapter outline

Use this chapter outline and the introductory page of the chapter in the Student Book to give students a mental roadmap for the chapter.

**6.1 The geography of technology** How the geographical spread of modern technology varies around the world

**6.2 Distribution of technology** Investigating the geographical distribution of technology in three areas: farming, telecommunications, and high-speed rail transport

**6.3 Access to technology – 1** Comparing how London and Dhaka are using different levels of technology to counteract the effects of flooding

**6.4 Access to technology – 2** Investigating how technology is being used to tackle the spread of HIV/AIDS, and whether technological breakthroughs are accessed equally

**6.5 The technological fix?** Learning about the link between economic development and technological innovation

**6.6 Technological leapfrogging** What technological leapfrogging is and how it can be used as a tool for development

**6.7 Impacts of technology** Investigating the impacts of technological innovation – in this case, genetically modified foods

**6.8 The effects of technology** Investigating how the effects of technology use are accounted for in some countries but not in others

**6.9 Contrasts in technology** How a range of technologies are being introduced to tackle water shortages in the Tigray region of Ethiopia

**6.10 The global fix** Investigating whether there are technological fixes to overcome global environmental issues such as global warming

**6.11 Technology and the future** The future opportunities for energy security and independence in Slovakia (Synoptic unit)

## About the topic

- This topic has two themes: technology, and the extent to which it can manage and solve some of the issues facing the world now.
- Access to technology is closely related to level of development. Just as development is uneven, so is the geography of technology.
- Many people rely on technology to solve problems, while others lack access to technology at basic levels.
- The use of technology has costs as well as benefits.
- Technology varies between large-scale top-down mega-projects and small-scale intermediate approaches.

## About the chapter

- This chapter begins by looking at the geographical spread of modern technology and the distribution of technology in farming, telecommunications, and high-speed rail transport.
- It looks at access to technology, comparing flood protection measures in Dhaka and London, and how technology is being used to tackle the spread of HIV/AIDS.
- It looks at the link between economic development and technological innovation, and how technological leapfrogging can be used as a means of development.
- The chapter examines the impacts of technological innovation by investigating GM foods – the costs and benefits – and how we deal with the externalities of increasing car use.
- A case study of Ethiopia illustrates how technology is being used at a range of scales to tackle water shortages.
- The chapter ends by looking to the future – whether we can find technological fixes to overcome problems such as global warming, and whether they can provide energy security and independence for Slovakia.

## Key vocabulary

There is no set list of words in the specification that students must know. However, examiners will use some or all of the following words in the examinations, and would expect students to know them and use them in their answers.

antiretroviral drugs
biomanufacturing
carbon capture technology
Digital Access Index (DAI)
digital blackout
DNA
environmental determinism
externalities
genetic modification
high-technology
information and communications technology (ICT)
intermediate technology
nanoparticle
pandemic
pasteurisation
patent
pervasive (in respect of technology)
Polluter Pays Principle
poverty line
technological leapfrogging
technology
technology-poor
technology-rich

**plus** examples of technology, e.g. solar technology, mobile phone technology, biomedical technology, genetic modification, iron fertilisation, $CO_2$ scrubber technology

The glossary at the end of this book contains many of these words and phrases. For students, the key word boxes in the chapter or the glossary at the end of the Student Book will help them with the meanings of all.

# 6.1 The geography of technology

## The unit in brief

This 2-page unit introduces the chapter on 'The technological fix?' It begins with the story of the day the Internet died for millions of people after an undersea cable was accidentally cut. It only took a few days for the cable to be fixed and the Internet to be restored, but it highlighted our dependence on modern technology in terms of the economic impact that this incident had.

The unit includes a Background box which explains the spatial variations in technology levels and what the technological fix is. It concludes with a map which shows that there is significant variation in ICT access between countries and continents.

## Key ideas

- We are highly dependent on modern technology.
- Technology means the development of knowledge, techniques, and systems which can be used to help solve problems and extend human capabilities.
- In geographical terms, technology can mean ways in which people innovate, change, or modify the natural environment to supply human needs or wants.
- Technology levels, and access to ICT, vary spatially.
- The technological fix relates to the expectations of people that continuing technological developments will help the world to tackle new problems as they arise.

## Unit outcomes

By the end of this unit most students should be able to:

- explain how the digital blackout of January 2008 could have had a major economic impact;
- define 'technology' in geographical terms;
- understand that technology levels vary spatially;
- explain how ICT can be said to be pervasive;
- give an example of a technological fix.

## Ideas for a starter

1. Use the 'What do you think?' on page 220 of the Student Book as a class discussion to start the unit. Ask students what it would mean to them if there was a digital blackout.
2. Brainstorm: What has technology got to do with geography? What is a technological fix?
3. Show the map from page 221 of the Student Book on the whiteboard without the caption. What do students think it shows? How is the data collected?

## Ideas for plenaries

1. If you did not use starter **1**, you could use it as a plenary.
2. Ask students to take two minutes with a partner to think up one interesting question about the geography of technology that is not covered in this unit.

# 6.2 Distribution of technology

## The unit in brief

This is a 6-page unit which investigates the geographical distribution of technology in three areas: farming, telecommunications, and high-speed rail transport. The first two pages look at farming and improvements in farming technology concentrating on genetic modification, and include three case studies of advances in farming technology in developing economies. The next two pages deal with Internet access (with a case study of access in East Africa), and the global distribution of mobile phones. The final spread looks at the development of high-speed rail transport in both Europe and Africa.

## Key ideas

- Advances in farming technology have been adopted in many countries.
- Since 1995, the Internet has grown to allow instant communication and information sharing.
- The rapid growth of the mobile phone industry has revolutionised communication, but not everyone has equal access to the technology.
- Rail infrastructure is an essential part of economic development and environmental planning.
- Europe has 3000 km of high-speed railway lines, with plans to build another 6000 km of lines by 2020.
- High-speed rail travel is due to begin operating in Morocco in 2013.

## Unit outcomes

By the end of this unit most students should be able to:

- explain how technological developments have improved farming methods since 1830;
- identify which continent is missing out on GM developments, and explain why this is so;
- list the advantages and disadvantages that the new Internet cable system might bring for East African Internet users;
- identify the patterns evident between ICT access and mobile phone distribution;
- describe the extent of the high-speed rail network in Europe, and the proposed high-speed rail lines for Africa;
- assess the extent to which they agree that some parts of the world remain disconnected from technological developments.

## Ideas for a starter

1. Use 'Over to you' activity **1** on page 227 of the Student Book as a starter. It will help to get students thinking about the technology that they use, and that others may not have access to.
2. Show either of the maps on pages 222 and 225 of the Student Book, or the table on page 224 on the whiteboard. Ask students to describe any patterns/what the data show.

## Ideas for plenaries

1. Get students to work in groups of three. They should take five minutes to write a paragraph on the distribution of technology in farming, telecommunications, or high-speed rail transport (one paragraph per student).
2. Play 'Just a minute'. The topic is 'The distribution of technology'. Students have one minute to talk on the topic without hesitation or repetition.

# Access to technology – 1

## The unit in brief

This 4-page unit compares how London and Dhaka are using different levels of technology to counteract the effects of flooding. London is a city of over 7 million people and has the sixth largest city economy in the world. The Thames Barrier became operational in 1982, but scientists now believe that a new flood defence scheme is needed to protect London from future flood surges.

The metropolitan area of Dhaka is home to over 12 million people and it is the commercial heart of Bangladesh's economy. Dhaka is vulnerable to flooding from monsoon rainfall and tidal surges. The unit investigates flood protection measures introduced since 1988, and the fact that despite the measures taken, Dhaka still floods.

## Key ideas

- The Thames Barrier became operational in 1982, and was expected to last until 2030.
- There are a number of reasons why London needs protecting.
- Scientists predict that the level of flood defence offered by the current Thames Barrier will not be enough to withstand future flood surges.
- Dhaka lies very close to sea level, making it vulnerable to flooding from monsoon rainfall and tidal surges.
- After disastrous flooding in 1988, the Dhaka Integrated Flood Protection Project was set up, with Phase II introduced after more flooding in 1998.
- Rapid urbanisation means that Dhaka's current flood protection measures can no longer cope.

## Unit outcomes

By the end of this unit most students should be able to:

- list the reasons why it was felt necessary to build a flood defence barrier in London;
- decide whether they think London will need a new flood defence system to cope with rising sea levels;
- describe the measures taken by the Dhaka Integrated Flood Protection Project and explain why some have not been successful;
- complete a table showing whether London is more geographically advantaged, or technologically advantaged, than Dhaka to enable flood protection measures to be successful.

## Ideas for a starter

1. Ask students if anyone has seen the Thames Barrier. If they have, ask them to describe it. What is it for? How does it work?
2. Show the photo of flooding in Dhaka in 2004 from page 231 of the Student Book on the whiteboard. Ask students where they think this place is. Why does it flood? How could it be prevented? Record students' ideas on a spider diagram.

## Ideas for plenaries

1. Use the 'What do you think?' on page 231 of the Student Book as a plenary to get students thinking about the links between global warming, wealth, and technology.
2. Brainstorm: What does the term environmental determinism mean? Once students have decided on the correct meaning, use 'On your own' activity **6** on page 231 as a class discussion.
3. If you used starter **2**, return to the spider diagram. How do students' ideas compare with the action taken by the Dhaka Integrated Flood Protection Project?

# 6.4 Access to technology – 2

## The unit in brief

This 4-page unit investigates the extent to which technology is being used to tackle the spread of HIV/AIDS, and whether technological breakthroughs for tackling the disease are accessed equally. The first two pages include data and facts about the HIV/AIDS pandemic as well as a Background box on HIV/AIDS. The final two pages look at a range of technological developments from different parts of the world that are being used to tackle the impacts of both HIV and AIDS.

## Key ideas

- According to UN estimates, there were about 33 million people living with HIV worldwide in 2007, and although the number of people is growing, the percentage has stabilised.
- Sub-Saharan Africa is the region most seriously affected by HIV/AIDS.
- Antiretroviral drug development continues to be the leading way of countering the effects of HIV.
- Recent technological developments are being used to tackle the impacts of both HIV and AIDS.

## Unit outcomes

By the end of this unit most students should be able to:

- describe the pattern of adult HIV infection rates at both global and regional scales;
- explain the reasons that have contributed to the concentration of the HIV epidemic in sub-Saharan Africa;
- explain the link between the availability of technology and the impact of HIV;
- complete a table to show how the new technological developments work, and whether they are available to all.

## Ideas for a starter

1. Introduce the topic of the use of technology to tackle the spread of HIV/AIDS by getting individual students to read out the five bullet points on page 232 of the Student Book. Do any of these figures surprise students?
2. Show either of the graphs, the map, or the photo of the HIV virus on pages 232 and 233 of the Student Book to kick off the unit.
3. Ask students why you are investigating HIV/AIDS in a geography lesson.

## Ideas for plenaries

1. Get students to work in pairs to come up with as many words as they can on how technology is being used to tackle the spread of HIV/AIDS.
2. How likely do students think it will be that there will be universal access to HIV/AIDS care by 2010? Debate.
3. Ask students to sum up what they have learned in this unit in 40 words or less, and to tell their partner.

# 6.5 The technological fix?

## The unit in brief

This 2-page unit looks at the link between economic development and technological innovation. Publication of *The Growth Report: Strategies for Sustained Growth and Inclusive Development* in 2008 highlighted the importance of technology, innovation, and higher education in achieving economic growth. The unit looks at those countries which have achieved fast and sustained growth since 1945, and the importance of innovation and participation rates in higher education in sustaining economic growth.

## Key ideas

- *The Growth Report* identified key factors for sustained high economic growth.
- Increased investment in innovation (which can be measured by the number of patents granted) and higher education is needed to ensure growth is sustained.
- Many countries may not have the ability to invent as quickly as they can adopt and learn new technologies.
- University education provides a highly skilled labour force which is needed for sustained economic growth.
- Industrialised countries have a responsibility to finance the expansion of Africa's university education to make up for Africa's 'brain drain'.

## Unit outcomes

By the end of this unit most students should be able to:

- list the key factors identified by *The Growth Report* for sustained high economic growth;
- explain how technology, innovation, and university education can help to promote development and growth;
- describe and compare the pattern of countries which have achieved fast and sustained economic growth with the pattern of top 10 countries for patent grants awarded and university enrolments;
- say how far they agree with the view that industrialised countries have a responsibility to finance the expansion of Africa's university education.

## Ideas for a starter

1. Show the table of the 13 countries which have achieved fast and sustained growth since 1945 from page 236 of the Student Book. Ask students what they think the table is about. If those countries have achieved growth, why have other countries not been able to?
2. Show the photo of James Dyson and the vacuum cleaner from page 237 of the Student Book. Ask: What is the link between the vacuum cleaner and geography? (The answer is technology, innovation, and economic development.)
3. Show the tables of the top 10 countries for registered patents and those with the highest university enrolments from the foot of page 236 of the Student Book. Ask: Why are these figures important? How are they collected?

## Ideas for plenaries

1. Use the 'What do you think?' on page 237 of the Student Book as a plenary.
2. Quick fire test:
   - What are the key factors for economic growth as identified by *The Growth Report*?
   - Name some of the countries which have achieved fast and sustained growth since 1945.
   - What is a patent, and why are they important?
   - Why is university education important to long-term economic development?

# 6.6 Technological leapfrogging

## The unit in brief

In this 4-page unit students find out what technological leapfrogging is, and how it can be used as a tool for development. It begins with an example from Afghanistan where the country's largest telephone provider has leapfrogged the need for a fixed-line infrastructure and moved into the wireless mobile sector. The rapid growth of mobile phone use has also been seen in other developing countries.

The unit also looks at the development of solar energy in India and Pakistan. In Pakistan, it is believed that providing electricity from alternative sources will lead to faster development, whilst in India a project training solar engineers provides a solution for cooking, lighting, education, agriculture, health, and income generation.

## Key ideas

- Afghanistan has leapfrogged the need for a fixed-line telephone infrastructure and moved into the wireless mobile sector.
- There is a range of benefits for developing countries in developing mobile phone networks and leapfrogging cabled networks of landlines.
- Technological leapfrogging may be the process by which the technology and development gap between countries is reduced.
- In Pakistan, it is believed that providing alternative sources of electricity to communities beyond the reach of grid-based power will lead to faster development.
- In India, a project training people to build, operate, and maintain their own solar-power systems provides a solution for cooking, lighting, education, agriculture, health, and income generation.

## Unit outcomes

By the end of this unit most students should be able to:

- describe the physical and human reasons that made the creation of a fixed landline infrastructure in Afghanistan difficult;
- list the benefits of developing mobile phone networks for developing countries;
- explain the process of technological leapfrogging;
- assess how far technological leapfrogging will help to reduce the technology gap between countries;
- explain how the provision of a reliable electricity supply in rural areas of Pakistan and India improves the quality of life there.

## Ideas for a starter

1. Show the photo of Abdul Wakil from page 238 of the Student Book on the whiteboard, and read out the first paragraph from the top of page 238. Ask students why they think there are very few fixed line telephones in Afghanistan. Can they imagine what life must have been like for Abdul Wakil before he got a mobile phone?
2. Brainstorm: What is technological leapfrogging?

## Ideas for plenaries

1. Ask students to write 'technological leapfrogging' in the middle of the page and create a mind map around it. How many ideas can students come up with in two minutes?
2. Ask students to identify three key ideas from this unit.

# 6.7 Impacts of technology

## The unit in brief

This 2-page unit investigates the impacts of technological innovation using the example of genetically modified (GM) foods. GM foods were first introduced in the USA in 1994, but public opinion rapidly turned against them. The unit looks at the costs and benefits of GM foods and includes a case study on Golden Rice. Golden Rice was developed in Switzerland to tackle the issue of Vitamin A deficiency in children across the world. The case study includes the concerns expressed by Friends of the Earth about Golden Rice.

## Key ideas

- GM foods were first introduced in the USA in 1994, but public opinion turned against them.
- GM foods were developed to feed an ever-growing population, and it is argued that they have many benefits.
- Organisations campaigning against GM crops believe that the long-term effects have not been fully tested.
- Golden Rice was developed in 1999 to tackle the issue of Vitamin A deficiency in children across the world.
- There are still concerns about Golden Rice.

## Unit outcomes

By the end of this unit most students should be able to:

- give reasons why GM foods cannot be considered to be a 'neutral' technology;
- complete a table to show the positive and negative impacts of GM foods on the economy, society, and the environment;
- explain why Golden Rice was developed;
- list the concerns about Golden Rice;
- explain whether they believe that GM foods will bring benefits to some of the poorest people in the world.

## Ideas for a starter

1. Put a traditional picnic basket on a desk, with some typical French food such as cheese, baguettes, etc. Read out the first two paragraphs on page 242 of the Student Book, or a version in your own words, to set the scene for this unit.
2. Ask students what they know about GM foods. What are the benefits? What are the costs? Record students' ideas as two spider diagrams.

## Ideas for plenaries

1. Spilt the class into two groups. One group are members of FNSEA, who support a scientific and highly productive approach to agriculture. The other group are GM protestors, who support a traditional, small-scale, GM-free approach. The groups should debate the issue of GM foods.
2. If you used starter **2**, return to the spider diagrams. Do students need to add or remove anything from their spider diagrams as a result of their work in this unit?

# 6.8 The effects of technology

## The unit in brief

In this 4-page unit students investigate the environmental costs, or externalities, of car use. The first two pages look at the 'Polluter Pays Principle' (PPP), and include a range of examples from Western Europe of how different countries are tackling the $CO_2$ emissions from cars. The second two pages look at the Indian car sales boom – from the manufacturing of the world's cheapest car (the Tata Nano) to the externalities of the Indian car boom. As European countries introduce measures to reduce $CO_2$ emissions from vehicles, it appears that growth in India could negate the global benefit from European schemes.

## Key ideas

- The PPP is the idea that those who cause pollution should bear the economic cost of the damage they are doing to the environment.
- Growing economies, such as China and India, are likely to treat the environment as a 'sink' to soak up emissions.
- Countries in Western Europe have introduced schemes to tackle the $CO_2$ emissions of cars.
- The Tata Nano was designed to target India's growing middle class.
- In 2004, India was the world's fourth largest $CO_2$-emitting country, but scientists are concerned that emissions will accelerate as a result of rapidly increasing car ownership.
- Growth in India could negate any benefit from European measures to reduce $CO_2$ emissions from vehicles.

## Unit outcomes

By the end of this unit most students should be able to:

- describe, with examples, the two different approaches of implementing the PPP;
- explain why growing economies are likely to treat the environment as a 'sink' to soak up emissions;
- explain to what extent European schemes will have an effect on global $CO_2$ emissions;
- understand why scientists and environmental critics are concerned about the introduction of the Tata Nano in India;
- assess whether, with increased car ownership in China and India, there is any point in European governments trying to reduce their own $CO_2$ emissions.

## Ideas for a starter

1 Tell students that in 2007 there were nearly 600 million cars on the road – nearly one for every 10 people. And that figure is likely to double by 2030. Ask: What measures can we take to reduce their environmental impact and tackle $CO_2$ emissions?

3 Ask: Who has heard of the Tata Nano? Show students the photo of the car on page 246 of the Student Book. Why is this car important?

## Ideas for plenaries

1 With books closed, ask: What are externalities? What is the Polluter Pays Principle? Give some examples of how European countries are tackling the $CO_2$ emissions of cars.

2 Use the 'What do you think?' on page 247 of the Student Book as a plenary.

3 Choose three or four students (give them advance warning) to act as Indian government ministers. They take 'hot seats' in front of the class. The class acts as reporters and fire sensible questions at them about what they are going to do about India's growing car use and $CO_2$ emissions.

# 6.9 Contrasts in technology

## The unit in brief

This is a 4-page unit in which students investigate how a range of technologies are being introduced to tackle water shortages in the Tigray region of Ethiopia. Ethiopia is considered to be one of the poorest countries in the world, yet since 2003 has seen a growth in its GDP equal to, or greater than, most other developing economies. A severe drought in 2008 created major problems and the World Bank has developed a Country Water Resources Assistance Strategy to help Ethiopia to provide year-round access to water. This has led to a range of projects from the high-tech Tekeze Dam to the low-tech 'harvest the rain' strategy.

## Key ideas

- Ethiopia is one of the poorest countries in the world, yet since 2003 has seen a growth in its GDP equal to, or greater than, most other developing economies.
- The World Bank has developed a Country Water Resources Assistance Strategy to help Ethiopia develop year-round access to water.
  - Low-tech solutions include 'harvesting the rain' to improve the water supply situation for Ethiopia's rural poor.
  - The Tekeze Dam is a high-tech solution. Water stored will be used for irrigation as well as electricity production.
  - Intermediate technology includes building small and medium-sized dams to provide water and food security for communities.

## Unit outcomes

By the end of this unit most students should be able to:

- understand the reasons why Ethiopia needs to overcome its water supply problems;
- complete a table to show the economic, social, and environmental impacts of the three technology schemes being used in the Tigray region of Ethiopia;
- decide whether the social and environmental consequences of large dam schemes are justifiable in order for countries to develop;
- explain how each of the three schemes aims to tackle water security for Ethiopians.

## Ideas for a starter

1 Have a bottle of Highland Spring mineral water and pass it around the class. As the bottle is going around, ask students: What is the link between bottled water, electricity, and Ethiopia? (The answer is that in 2008 drought struck in Ethiopia. That year there was a lack of bottled water (including Highland Spring) which was due to a lack of power supply. One of the consequences of the drought was that HEP dams were not able to operate at full capacity and water companies had to cut back on production due to lack of electricity supply.)

2 Introduce students to Tadesse Desta from the foot of page 249 of the Student Book. What low-tech solutions could he adopt to improve his water supply situation?

## Ideas for plenaries

1 Use 'On your own' activity **4** on page 251 of the Student Book as a plenary and debate the World Bank's view that 'Ethiopia is poor because it doesn't use its enormous water potential'. Students can write up the views expressed later.

2 Students could work in threes to each write a paragraph about Ethiopia's low-tech, high-tech, and intermediate technology solutions to its water supply problems.

3 Play 'Just a minute'. The topic is 'Water shortages in Ethiopia'. Students have a minute to talk on the topic without hesitation or repetition.

# 6.10 The global fix

## The unit in brief

In this 4-page unit students investigate whether there are technological fixes to overcome global environmental issues, such as global warming. In 2007, Sir Richard Branson and Al Gore launched the Virgin Earth Challenge prize offering $25 million for the best idea for removing at least one billion tons of $CO_2$ from the atmosphere each year. Since the announcement of that prize scientists have begun to suggest possible solutions to global warming, some of which are included in this unit. Other ideas include finding a man-made way of replicating a natural process – carbon capture technology – using artificial trees to remove $CO_2$ from the atmosphere.

## Key ideas

- Boserup argues that rapid increases in population are accompanied by rapid technological change – hence technology can provide a global fix for global issues.
- There are a range of possible technological solutions to global warming, including giant sunshades, iron fertilisation of the oceans, and artificial volcanoes.
- The idea for artificial carbon capture trees is based on finding a man-made way to replicate a natural process.
- Carbon capture technology involves collecting the carbon and storing it.

## Unit outcomes

By the end of this unit most students should be able to:

- explain Boserup's view about technological change;
- produce a table to assess the costs and benefits of the technological solutions to global warming included in this unit;
- describe how carbon capture technology works;
- assess whether technology to remove $CO_2$ from the atmosphere is the technological fix to the problem of global warming.

## Ideas for a starter

1. Tell students that Sir Richard Branson and Al Gore have launched a prize offering a $25 million reward for the best idea to remove one billion tons of $CO_2$ from the atmosphere each year. Can they come up with any technological solutions that would be worthy of the prize?
2. Ask: What is carbon capture technology? How does it work?

## Ideas for plenaries

1. Use the 'What do you think?' on page 255 of the Student Book as a plenary.
2. Read aloud Dr Lackner's view at the top of page 255, and then use 'Over to you' activity **3** on page 255 as a plenary.
3. What other technological solutions can students think of to tackle global warming?

# 6.11 Technology and the future

## The unit in brief

This is the synoptic unit for the chapter on 'The technological fix?' It is a 4-page unit in which students look at opportunities for energy security in Eastern Europe, focusing on Slovakia. Following the break up of the USSR, many countries were still economically tied to Russia. In Slovakia's case, it depends on Russia for 82% of its oil, and nuclear power provides 53% of its electricity – a hang over from the Cold War era. The unit provides students with a range of resources which look at Slovakia's energy security and possibilities for the future. It ends with a synoptic question to give students an idea of what examiners could ask in the exam.

## Key ideas

- Following the break up of the USSR, many Eastern European countries were still economically tied to Russia.
- Slovakia is dependent on Russia for 82% of its oil.
- 53% of Slovakia's electricity comes from nuclear power.
- The contribution of renewable energy to Slovakia's energy supply is minimal.
- Slovakia's energy policy review in 2005 included recommendations to:
  - ensure energy security;
  - reduce dependence on nuclear power;
  - increase the use of renewable energy;
  - tackle energy pollution.

## Unit outcomes

By the end of this unit most students should be able to:

- explain the factors that led to Slovakia's energy supply being dependent on imported oil and technology;
- assess the sustainability of Slovakia's current energy sources;
- assess whether renewable energy resources will provide a technological fix to give Slovakia energy security and independence.

## Ideas for a starter

1. Show the table of the top 10 users of Russian oil from page 256 of the Student Book on the whiteboard. Ask: Is Slovakia right to be concerned about its dependence on Russian oil?
2. Show the graph of Slovakian energy production from page 257 of the Student Book on the whiteboard. What problems might Slovakia's dependence on nuclear power cause?

## Ideas for plenaries

1. Students could work in pairs to plan an answer to one part of the synoptic question on page 259 of the Student Book.
2. Give students a blank map of Slovakia. Get them to annotate it with ideas on Slovakia's energy security.
3. Ask students to identify three key things they have learned in this unit.

# 7 Tectonic activity and hazards

Chapter 7 in the Student Book is a sample study, which along with Chapters 8–12 is designed to help students select which Option to study. Students can, and should, use these chapters in their research, but they do not form complete courses. For extra resources on this Option, refer to *A2 Geography for Edexcel Activities & Planning OxBox CD-ROM*.

## About the Option

- Tectonic activity generates a wide range of natural hazards, caused by plate tectonics.
- Tectonics is a key landscape-forming process, producing distinctive landforms in active regions, ranging from minor features such as scarps to vast rift valleys and shield volcanoes.
- Tectonic hazards cause risk to people and their possessions, depending upon their vulnerability, and the magnitude and frequency of the event.
- Risk varies due to factors including level of development, preparedness, and education.
- Hazard impacts may be short-term or long-term.
- People respond to hazard risk in different ways, depending upon their knowledge, technology, and financial resources.

## Introducing the Option

The movement of the Earth's tectonic plates can be hazardous for human activity. Volcanic eruptions, earthquakes, and tsunami often grab the news headlines when many lives are lost. The short-, medium-, and long-term impacts of these tectonic events vary in relation to the intensity and frequency of the event, and the nature of the location affected. Levels of economic development, methods of prediction and preparation, and population densities often determine the severity of hazardous events. How people cope and recover depends on the scale and nature of the event.

- Some events are dramatic. The 2004 Boxing Day tsunami, which was caused by an earthquake with a magnitude of 9 to 9.3 under the Indian Ocean (near the coast of Sumatra) killed over 150 000 people. Although the earthquake was localised, the effects of the surge of water were felt over a vast area and reconstruction will take years.
- Some events are small scale and go unnoticed by the rest of the world. Minor earth tremors and volcanic eruptions occur throughout the year in Iceland. They usually only cause short-term disturbance and help to generate valuable tourist revenues as people deliberately visit the country to watch the geysers and bathe in geothermally warmed pool, while Icelanders enjoy cheap central heating and year-round salad crops and fruit grown in glass houses heated by the geothermal activity.
- Tectonic activity can also create distinctive landscapes, like the Great African Rift Valley, fault planes, and mountains – as well as trigger other geomorphic hazards, such as landslides, floods, and climatic change.

## Using Chapter 7

Chapter 7 contains six pages of resources.

- Page 264 introduces the topic with a case study of Montserrat.
  - Following two years of small eruptions between 1995 and 1997, the Chances Peak volcano fully erupted in June 1997.
  - Although Plymouth (Montserrat's capital) was destroyed, advance preparation and warnings saved many lives.
  - The government's four-year Sustainable Development Plan, launched in 2003, aimed to restore confidence, rebuild infrastructure, and the economy.
- Pages 265–6 investigate the issue further using resources. Students are guided, but will need to think about each resource carefully. The resources look at:
  - Montserrat's revival
  - funding Montserrat's recovery
  - how the risks of living on Montserrat are being managed.
- Page 267 provides background information about tectonic hazards and risks.
  - Disaster was avoided in Montserrat as two-thirds of the island was declared an exclusion zone before the 1997 eruption.
  - The people of Montserrat still face an uncertain future and the island's economy has not recovered.
  - People living in naturally volatile areas are exposed to increasing risks and are considered vulnerable. When vulnerability and hazards coincide, there's a high risk of disaster.
  - As global population grows more people are living in danger zones.
- Pages 268–9 investigate the issue in other areas of the world, again using resources. These look at the impacts of earthquakes in areas with differing population densities – the Chinese earthquake in May 2008, and the effects of earthquakes in Japan.
- Page 270 provides activities, useful websites for further research, films, books, and music on the theme, and examples of the kinds of questions that students will meet in the exam.

Building up a file of tectonic hazards will help students to understand issues about their impacts and management. Useful tectonic events include:

- in 1963, Icelanders were able to witness the birth of Surtsey, a new volcanic island across the mid-Atlantic constructive plate boundary;
- the Pakistan earthquake in Balakot which initially killed 19 000 in October 2005, with a further 60 000 dying in the aftermath following a harsh winter in an area of poverty;
- the largest earthquake in the UK for 25 years on 27 February 2008 (the epicentre of the 5.2 magnitude earthquake was near Market Rasen in Lincolnshire);
- 15 000 Colombians were forced to evacuate as the Nevado del Huila Volcano began to erupt in April 2008; between May and June 2008, the villagers around the Chaitén Volcano in Chile faced outpourings of ash.

# 8 Cold environments: landscapes and change

Chapter 8 in the Student Book is a sample study, which along with Chapters 7 and 9–12 is designed to help students select which Option to study. Students can, and should, use these chapters in their research, but they do not form complete courses. For extra resources on this Option, refer to *A2 Geography for Edexcel Activities & Planning OxBox CD-ROM*.

## About the Option

- Cold environments include glacial uplands, high-latitude ice-bound regions, and periglacial areas.
- The distribution of these regions has changed during the Quaternary period, and continues to change now.
- Climate determines the location of cold environments, and many landscape features in the UK and elsewhere result from past and present geomorphological processes that operate in cold environments.
- Cold environments present people with both challenges (hazards) and opportunities (the resources found in cold regions).
- Cold environments are increasingly under threat from human actions and require management and protection.

## Introducing the Option

We know that the Earth's climate is currently changing; scientific research shows that it often changes. In the last 2.6 million years of geological time (known as the Quaternary period), major shifts of global climate have occurred several times. Each time, ice sheets extended into temperate latitudes – bringing glaciers to upland areas, altering the landscape, changing sea level, and shifting plant and animal habitats. This all happened in a period shorter than 0.05% of the Earth's total history!

Cold environments (i.e. those with temperatures permanently <0°C) give rise to distinctive landscapes, which not only display the power and presence of ice but also restrict human activity. Beneath and beyond the ice masses, natural systems reflect changes in global climate, which can be:

- rapid – like the collapsing Hornbreen glacier in Svalbard, part of which crashed into the Arctic Ocean in August 2007, causing a huge wave which injured 17 British tourists aboard a nearby sightseeing ship;
- slow – such as the gradual retreat of glaciers and ice sheets down the valley of the Grossglockner (Austria), or in most of southern Iceland, uncovering glacial deposits;
- preserved – as landscape features from past ice ages, such as the 3km-deep Sognefjord (Norway), or the more modest lumps of Lake District rock (called erratics) in Stoke-on-Trent.

## Using Chapter 8

Chapter 8 contains six pages of resources.

- Page 272 introduces the topic with a case study about the Arctic region.
  - The Arctic Ocean ice sheet and permanent year-round ice are thinning.
  - As ice thins, economic opportunities emerge: the North-West passage was entirely navigable in 2007, and raised tensions over access and ownership; exploitation of oil and gas below the Arctic Ocean will become easier, but Arctic countries are in dispute over territorial and exploitation rights.
- Pages 273–4 investigate the issue further using resources. Students will be guided, but will need to think about each resource carefully. The resources look at:
  - the Arctic's emerging place in the global economy;
  - the challenge of development in cold environments;
  - activities that have already reached the Arctic.
- Page 275 provides background information about the distinctiveness of cold environments.
  - The Arctic region of the twenty-first century is a fraction of the size of the cold environment that existed 18 000 years ago.
  - Thawing permafrost leads to a number of problems.
  - By studying present-day processes operating in Canada's tundra, or Iceland's glaciers, it is possible to learn how many of Britain's landscapes were created in the same way.
- Pages 276–7 investigate the impact of past processes in Britain, again using resources. They consider the effects of glacial and periglacial environments on landscapes in Britain.
- Page 278 provides activities, useful websites for further research, films, books, and music on this theme, and examples of the kinds of questions that students will meet in the exam.

Building up a file on cold environments will help students to understand the processes and changes that have occurred and do occur in these parts of the world. Useful examples of distinctive landscapes and challenges to human activity include:

- the dry valleys and development of coastal features on the Isle of Purbeck around Lulworth Cove;
- contemporary periglacial conditions in the Cairngorms and Scottish Highlands;
- the challenges involved in the construction and maintenance of the Trans-Alaskan oil pipeline, including the impact on local ecosystems and people;
- the growing threats of modernisation on the Saami reindeer herders of Lapland and the Inuit of northern Canada;
- the extension of the 'pleasure periphery' to Antarctica, the Falkland Islands and northern Finland.

# 9 Life on the margins: the food supply problem

Chapter 9 in the Student Book is a sample study, which along with Chapters 7, 8, and 10–12 is designed to help students select which Option to study. Students can, and should, use these chapters in their research, but they do not form complete courses. For extra resources on this Option, refer to *A2 Geography for Edexcel Activities & Planning OxBox CD-ROM*.

## About the Option

- Significant numbers of people live in a state of food insecurity, while others consume more than their share of global resources. This Option explores this inequality, particularly in regions where food production is a challenge.
- The 'margins' might be traditional areas of famine, or rapidly urbanising areas where food is scarce and malnutrition threatens.
- The causes of food insecurity range from physical (e.g. land degradation and desertification) to human (over-exploitation, population pressure, political processes).
- Increasing food supply forms one of the Millennium Development Goals.
- The ways to achieve increasing food supply are contentious, e.g. reforming trade systems, hi-tech farming, intermediate technology.

## Introducing the Option

Maintaining an adequate food supply for a rising global population remains a major challenge. For people living in poverty, or on marginal land where crops frequently fail, or those confronted by surges in commodity prices, the realities are stark. This Option focuses on the problems of food insecurity and the attempts being made to ensure food supplies for everyone.

- The intensification of agriculture increased crop yields by developing new high-yield, disease-resistant varieties of crops, and using machinery, fertilisers and pesticides to create an artificial ecosystem to support increasing human needs. These changes became known as the Green Revolution and transferred western technology to developing countries in the 1960s and 1970s to help them feed themselves. It had most impact in South and East Asia and South America, but less impact in sub-Saharan Africa.
- Major international businesses now dominate global food supply chains, having acquired premium land for plantations, ranches, and crop production at the expense of local smallholders.
- Global talks have tried to establish free trade by opening markets up and reducing unfair tariffs and subsidies that support American and European farmers.

## Using Chapter 9

Chapter 9 contains six pages of resources.

- Page 280 introduces the topic with a case study about Kalahandi in India.
  - Kalahandi, in Orissa state, is remote and marginalised.
  - Rainfall is unreliable; droughts and famine are common.
  - Between 1998 and 2003, rice production in Kalahandi exceeded local needs, but farmers could not afford to buy the rice grown for their landlords.
  - The 'Kalahandi syndrome' happens globally. 800 million people worldwide cannot afford to buy their recommended daily calorie intake.
- Pages 281–2 investigate the issue further using resources. Students will be guided, but will need to think about each source carefully. The resources look at:
  - the impacts of globalisation on rural India;
  - the impacts many rural Indian communities face from spiralling debt;
  - diversification to increase food purchasing power for the poor.
- Page 283 provides background information about the global food crisis and risks.
  - The World Bank predicts that global demand for food will double by 2030.
  - In 2008, over 800 million people worldwide were classified as undernourished.
  - There is a range of factors influencing the growing global food crisis.
- Pages 284–5 investigate the issue in other areas of the world, again using resources. These look at the emerging global food crisis and suggest that the problems are often in the solutions.
- Page 286 provides activities, useful websites for further research, books to read on this theme, and examples of the kinds of questions that students will meet in the exam.

Building up a file illustrating aspects of food supply around the world can strengthen students' understanding of this topic. Useful examples include:

- Ethiopia – the focus of the 1980s African food crises, where exports of coffee, nuts, and other cash crops continued as millions starved, civil wars raged, and droughts made matters worse. After decades of aid, the food problem remains and exports are higher than ever;
- Mexico – where the biotech lobby are fighting off farmers' organisations, which fear that GM crops threaten native corn varieties, livelihoods, and the nation's food sovereignty by increasing dependence on transnational seed companies;
- post-1982 Structural Adjustment Policies advocated by the World Bank to help indebted countries 'earn more from exports to pay back their loans'.

# 10 The world of cultural diversity

Chapter 10 in the Student Book is a sample study, which along with Chapters 7–9, 11, and 12 is designed to help students select which Option to study. Students can, and should, use these chapters in their research, but they do not form complete courses. For extra resources on this Option, refer to *A2 Geography for Edexcel Activities & Planning OxBox CD-ROM*.

## About the Option

- Culture is a complex concept. It varies spatially; some areas are homogenous, while others are hugely diverse.
- Large urban areas are often most diverse, reflected in their populations, services, and built environments.
- Attitudes to cultural diversity differ, both personally and politically. Some see globalisation as driving towards a global culture, aided by media TNCs and communications technology.
- Culture remains complex; local cultures survive and new cultures are born.
- Culture determines economic and environmental attitudes to consumption, conservation, exploitation, and protection.
- The conflict between consumer capitalism and the global environment is not easy to resolve.

## Introducing the Option

Modern communications now allow us to measure geographical distance in time as well as kilometres. Thanks to air travel, e-mail, mobile phones, and social networking websites, people from different countries with different values and experiences come into contact with each other on a regular basis. Information technology spreads ideas faster than ever, and now affects how people consume goods and services. Music, TV programmes, and films made by large media TNCs combine with migrations of people to create cultural mixes – or hybrids. This Option focuses on how economic development can lead to hybrid cultures, and examines whether a single 'global culture' could replace cultural diversity.

- Some cultural shifts reflect historical invasions, like those of Central and South America where Spanish and other European invaders superimposed Christianity and a new economic system on their colonies, in place of the existing ancient cultures.
- Other cultural changes are enforced when governments and/or international institutions require local communities to give up their traditional ways of life. For example, in the case of rural re-settlement programmes in Brazil where the IMF urged increased cash crop production instead of traditional subsistence farming.
- And some cultural changes are simply the result of exposure to the global media and instant communications, with fashion, music, films, and language reflecting a globalised consumerist society.

## Using Chapter 10

Chapter 10 contains six pages of resources.

- Page 288 introduces the topic with a case study of the Orang Asli peoples of Malaysia.
  - Malaysia's Orang Asli Semai people are descendents of the earliest people to live on the peninsular, and their relative remoteness has, until recently, protected them from change.
  - The economy of the central highlands area has developed rapidly recently.
  - The culture of the Orang Asli is under real pressure as the Malaysian government tries to integrate them into mainstream Malay society.
- Pages 289–90 investigate the issue further using resources. Students will be guided, but will need to think about each resource carefully. The resources look at:
  - traditional Orang Asli culture;
  - the scale of changes that have occurred in recent years.
- Page 291 provides background information on forces of cultural change.
  - Economic development changes landscapes and how people see the world.
  - Globalisation is a threat to cultural diversity.
  - Are we part of a global village, or an urban world?
- Pages 292–3 investigate the issue in other areas of the world, again using resources. These look at the erosion of cultural diversity in wealthy nations, and explore examples where attempts are made to protect cultural diversity.
- Page 294 provides activities, useful websites for further research, films, books, and music on the theme, and examples of the kinds of questions that students will meet in the exam.

Building up a file to illustrate aspects of cultural change around the world can strengthen students' understanding of this topic. Useful examples include:

- Bhutan – a small Buddhist kingdom in the Himalayas that has long avoided contact with the outside world, but since the opening of an international airport in 1983 – and the arrival of television in 1998 – is facing up to stark choices for its future;
- the Yanomani Indians of the Brazilian rainforest who were only 'discovered' in the 1970s and have been fighting to preserve their way of life since;
- Saipan – a Pacific island where 50% of the population now consist of immigrant workers from China, the Philippines, Bangladesh, Sri Lanka, and Thailand.

# 11 Pollution and human health at risk

Chapter 11 in the Student Book is a sample study, which along with chapters 7–10 and 12 is designed to help students select which Option to study. Students can, and should, use these chapters in their research, but they do not form complete courses. For extra resources on this Option, refer to *A2 Geography for Edexcel Activities & Planning OxBox CD-ROM*.

## About the Option

- Human health is a key concern. Personally, health impacts upon quality of life, but it also affects wider economic development and presents challenges regarding the spread of disease.
- Health risk is strongly linked to levels of economic development, either in the form of transmissible disease or environmental pollution.
- The spread of risk follows geographical patterns and features.
- Pollution is a key risk especially in countries where rapid economic development takes precedence over environmental and health concerns.
- A wide range of strategies can be adopted to manage pollution and health; some problems are harder to manage than others and require long-term strategies, and economic and lifestyle changes.
- Increasingly management is international in nature, reflecting an interconnected world.

## Introducing the Option

Pollution incidents vary. Most are linked to economic and human activity, for example, from a factory, traffic, etc. Although media reports about many incidents focus on the effects of pollution on ecosystems, like oil spills from tankers killing wildlife, this Option focuses on pollution and human health. Incidents and their effects on health vary.

- Some are dramatic. In Bhopal, a city in India, toxic gas escaped from a chemical plant in November 1984, killing thousands at the time and leaving a legacy of long-term sickness in the city.
- Some are one-off, local incidents. In Camelford, Cornwall, a water-filtering plant caused a spill of sulfuric acid into the River Camel. This had serious impacts on ecology and water quality, and the health of hundreds of people who drank the water.
- Some pollution is long-term. Over time, prolonged exposure damages health, such as the example of asbestos in this Option.

## Using Chapter 11

Chapter 11 contains six pages of resources.

- Page 296 introduces the topic of asbestos, with a case study of Wittenoom in Western Australia.
  - Wittenoom boomed in the 1950s due to mining crocidolite (blue asbestos).
  - Exposure to asbestos causes asbestosis and the cancer, mesothelioma.
  - The mine closed in 1966, and in the late 1970s the state government tried to close the town – a few residents still remain.
- Pages 297–8 investigate the issue further using source materials. Students will be guided, but will need to think about each source carefully. The resources look at:
  - Wittenoom in 2007;
  - the risks of being in Wittenoom for different groups of people;
  - the companies who mined and processed asbestos products.
- Page 299 provides background information about asbestos and risk.
  - Asbestos is a naturally occurring mineral which was widely used between 1950 and 1980.
  - People most at risk of asbestosis and mesothelioma are men, born between 1943 and 1948, who worked in mining and milling asbestos, and in building trades.
- Pages 300–1 investigate the issue in other areas of the world, again using source materials. These look at the implications of asbestos use in other countries, and how people cope with the risks of asbestos use.
- Page 302 provides activities, useful websites for further research, films, books, and music on this theme, and examples of the kinds of questions that students will meet in the exam.

Building up a file of pollution incidents will help students to understand the issues about pollution and human health. Other useful incidents include:

- London smog in 1952, where a lethal combination of weather conditions and smoke pollution caused deaths from bronchitis and other lung infections;
- Bhopal in India in 1984, where a faulty pipe in a chemical plant ruptured, causing toxic gas to escape, poisoning thousands of people;
- Chernobyl in 1986, where human error caused an explosion in a nuclear power station in Chernobyl, Ukraine, that spread pollution across Europe.

# 12 Consuming the rural landscape: leisure and tourism

Chapter 12 in the Student Book is a sample study, which along with Chapters 7–11 is designed to help students select which Option to study. Students can, and should, use these chapters in their research, but they do not form complete courses. For extra resources on this Option, refer to *A2 Geography for Edexcel Activities & Planning OxBox CD-ROM*.

## About the Option

- Increasingly landscapes are undergoing a structural shift from production to consumption, i.e. from primary production towards tourism and leisure.
- This shift affects all rural landscapes from the accessible rural-urban fringe to remote regions. Few areas remain untouched by leisure and tourism.
- Consumption puts pressure on often fragile rural landscapes, and is a threat that requires careful management.
- Reconciling the demands of consumers with the need to protect rural landscapes is a key challenge, which can be tackled in a wide variety of ways from preservation to ecotourism.

## Introducing the Option

Rural areas are being consumed by a wave of leisure and tourism, challenging local economies once devoted to food production or dependent on other primary activities. This Option focuses on the challenges posed by this ever-expanding sector of the global economy. Rural areas, which previously produced for consumption, are now being consumed themselves. The challenges vary according to the location.

- In the rural-urban fringes, short spells of leisure time are spent on a wide range of formal and informal activities that consume the landscape, such as golf courses, sports arenas, country parks, theme parks, walking and 'horsiculture', which are forcing people to reconsider how the countryside appears and is used.
- The Spanish Costas and Australian Gold Coast went from boom to almost bust as their carrying capacities were exceeded by over-development, degrading the natural appeal which first attracted the tourists and putting a strain on natural resources such as water supplies.
- Some remote wilderness areas, such as Antarctica, are now being developed for tourism. The litter trails of the Himalayas and Machu Picchu show what could happen here if uncontrolled tourism is allowed.

## Using Chapter 12

Chapter 12 contains six pages of resources.

- Page 304 introduces the topic with a case study of ecotourism in Peru.
  - Posada Amazonas in Peru is an eco-lodge owned by the local Ese'eja community and managed in partnership with a local Peruvian company.
  - The project offers tourists an insight into the rainforest ecosystem and boosts the local economy.
  - By making culture and nature commodities, indigenous people's values are changed.
  - Ecotourism offers an alternative route to development without destroying the very things that visitors go to see.
- Pages 305–6 investigate the issue further using resources. Students will be guided, but will need to think about each resource carefully. The resources look at:
  - the effects of Posada Amazonas;
  - a definition of ecotourism;
  - the risks of opening up remote rainforest areas to development.
- Page 307 provides background information on the pleasure periphery and models of tourism development.
  - The furthest distance that tourists travel is known as the 'pleasure periphery'.
  - Development of the tourist industry changes the character of a destination.
  - Tourists can exceed both the physical and social carrying capacity of a destination.
  - Designating places as National Parks and Heritage Sites, etc. may not protect them from the pressures associated with leisure and tourism.
- Pages 308–9 investigate the issue in other areas of the world, again using resources. These look at the wider impacts of leisure and tourism, and explore examples of managing leisure landscapes.
- Page 310 provides activities, useful websites for further research, films, books, and music on this theme, and examples of the kinds of questions that students will meet in the exam.

Building up a file to illustrate the impacts of the leisure and tourism industry around the world can strengthen students' understanding of this topic. Other useful examples include:

- the development of specialist leisure activities like safaris, wildlife, adventure, and active pursuits in Kenya, Zimbabwe, and Botswana
- opening up cold environments like Antarctica, Iceland, and Svalbard
- the increasing numbers of second homes in British National Parks
- the impacts of 'all-inclusive packages' on the Caribbean islands
- the relocation of weekly leisure activities to the urban-rural fringes of American and British cities.

# 13 Contested planet – synoptic investigation

## Using Chapter 13

In Chapter 13 students will develop an overview of Unit 3. Although the chapter focuses on Dubai, it also links the six Core topics of the A2 Specification and addresses the question of 'managing the contested planet'. It contains seven pages of resources:

- Pages 311–12 outline the scale of change taking place in Dubai.
    - Dubai is planning for a future without oil and is in the process of re-inventing itself as a 'global city'.
    - Dubai is redefining the concept of sustainable development, developing for the future with wealth creation as the driving force.
    - Dubai's primary objectives are to attain the status of a developed economy, and to build the first non-oil economy in the region.
    - Some critics are suggesting that Dubai is developing too quickly.
    - Dubai wants to continue its sustained economic growth, and increase its GDP and per capita income.
- Page 313 provides background information about the United Arab Emirates, of which Dubai is a major part.
    - Dubai has a number of geographical advantages.
    - Dubai has political and economic stability.
    - Dubai has a well-developed infrastructure and service sector.
- Pages 314–15 use resources to investigate the ways in which Dubai is changing. Students will need to think about each resource carefully. The resources look at:
    - Dubai's population surge;
    - the way in which the future is being built;
    - the scale of current developments in Dubai.
- Pages 316–17 question the sustainability of Dubai's development strategies, again using resources. The resources look at:
    - increasing pressures on Dubai's water supplies;
    - Dubai's sustainability credentials.
- Page 318 provides an additional resource which suggests that Dubai has no society of its own, together with useful websites for further research, and examples of the kinds of questions that students will meet in the exam.

This chapter gives students an indication of what it means to be synoptic – bringing several geographical issues and threads together in order to understand the challenges of managing a contested planet. Although it provides useful case-study material, it will not tell them everything! Students will need to consider the links between each of the themes and find evidence elsewhere in the Student Book, and from their own studies, to illustrate contrasting and similar situations.

The diagram opposite shows how this case study links to other themes and topics in the AS and A2 course.

For further guidance on synoptic investigations, see pages 24–27 in this Teacher's Book.

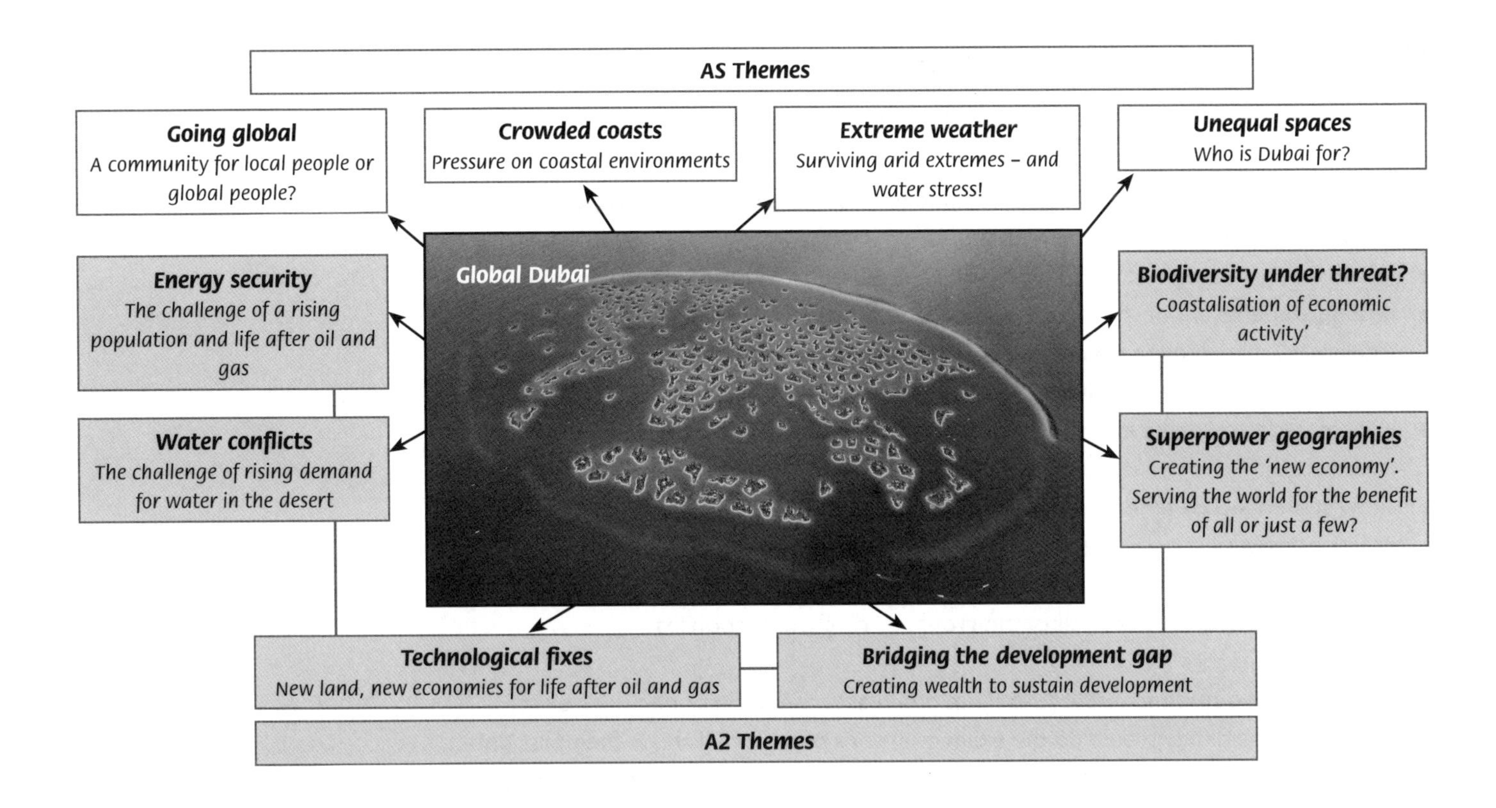


## Teaching the synoptic investigation

Depending upon when you want to use the synoptic materials on Dubai, there are two possible approaches.

### *Using Chapter 13 as a teaching unit*

You can use the materials exactly as you might any other lesson. Remember that the key to students understanding synoptic materials is not you teaching all they need to know, but engaging them with the resources. In this approach, the resources are analysed, in a class lesson in which you:

- introduce the materials;
- divide students into small working groups (2–3 is ideal) to consider a selection of the resources;
- allow students time to interpret these, and discuss them;
- get feedback from groups and compare responses.

Following an examination of all the resources over a few lessons, students could be given time to draft rough answers to the questions (group work is ideal for planning this kind of response), and write them up for homework.

This approach teaches particular techniques where you think appropriate – e.g. cost-benefit analysis – while students are learning about the resources.

It makes a good approach if you combine teaching this synoptic unit on Dubai integrated with Water, as this is (depending on your order of teaching) one of the first topics that students will meet. It also provides students with an expectation of what the synoptic unit may be like at an early stage, and reduces anxiety on their part.

## Using Chapter 13 for exam practice

If you are teaching the synoptic materials in Chapter 13 close to the time of release of the exam resources, you may decide to use them for practice. In this approach, the resources are used in preparing students, who then come to class where the focus is upon the examination questions. In this case, you would:

- introduce the materials in class briefly, to raise awareness of what is provided;
- remind students about the format of this part of the exam, with three questions totalling 40 marks, over a recommended time of 70 minutes;
- set private study time for fuller study of the resources;
- in class, assign students into small working groups of 2–3 to prepare notes on the three questions;
- allow students class time to prepare either plans or rough notes for their answers – or practice answers;
- get feedback from different groups and compare student responses. Ask students to discuss their plans with the rest of the group, or read out responses.

Following this, give students a full, 70-minute timed exercise to be done under exam conditions, in which they would do the exam questions prepared. If this is their first timed exercise, allow them to use notes; if it is the second, use it as a timed exercise without notes. Much will depend upon the ways in which you have used the synoptic enquiries in previous chapters.

# Exams: how to be successful

## Preparing students for exams

No matter how much your students enjoy Geography, preparation for the exam is essential. There is much to be gained by adding to your own and student enthusiasm for the subject, with a healthy dose of examination 'schooling' and preparation.

A2 is very different in its assessment from AS. Extended writing is demanded at all levels, culminating in a single in-depth 70-mark essay written for Unit 4. For Unit 3, examination success demands an ability to write in a focused way – but in depth.

## How students are assessed

There are two exams for A2 Geography; there is no coursework. The table shows the nature of the exams for Units 3 and 4.

| Unit | Assessment information | Marks and method of marking |
|---|---|---|
| Unit 3 Contested planet | This exam is 2 hours and 30 minutes long, and consists of two sections. It includes a pre-release resource booklet (about 4 pages long), designed to draw all the themes of the course together – known as the **synoptic** element.<br><br>The synoptic resources are **pre-released** as advance information **four weeks** before the exam. These will be about one of the six topics in Unit 3, and will form the basis of questions in **Section B** of the exam. Note that this topic will not be examined again in **Section A**.<br><br>There is no restriction on the use of the resources, but students must not take these into the exam. They will receive a fresh copy in the exam resource booklet.<br><br>The exam consists of two sections, A and B:<br>• **Section A** consists of **five** questions, each about the topics in Unit 3. Students select and answer **two**. Each question has two parts, worth 25 marks in total. Part (a) will normally be about a resource such as a map, diagram, or data. Part (b) will normally require students to answer based on their knowledge and understanding. Students are explicitly recommended to spend 80 minutes on this section, i.e. 2 exam answers x 40 minutes.<br>• **Section B** is the synoptic investigation based on the resource booklet. It consists of **one compulsory question** in three parts worth 40 marks in total. Students are explicitly recommended to spend 70 minutes on this section. | There are 90 marks in total for this exam, which are then converted to a UMS mark out of 120. All questions are level marked. |
| Unit 4 Geographical research | This exam is 1 hour and 30 minutes long. Whichever Option is chosen for Unit 4, there are four themes within it. **Four weeks** before the exam, centres will be notified about which **two** of these themes will be assessed in the exam. The exam question will be a single essay question drawn from the two themes.<br><br>In the exam, there are six questions in total – one for each of the six Options. Students must select and answer the one question relating to the option that they have studied. They need to write a single long essay in which they use and apply the results of their research.<br><br>Like Unit 3, students may not take any research material into the exam. | There are 70 marks for this exam, which are then converted to a UMS mark out of 80. Questions are level marked. |

## The exam questions

The table above explains the types of questions in each exam.
Three examples follow.

***Sample question for Unit 3 Section A***

**2** Study Figure 2.

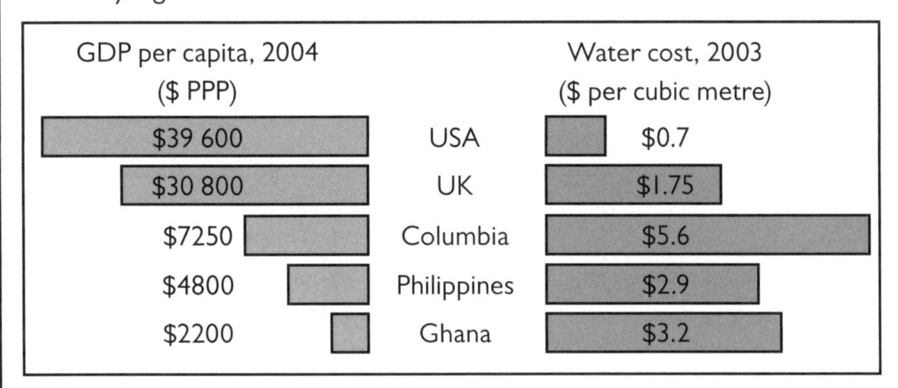


*Figure 2: Per capita GDP compared to the cost of water for five countries.*

**a** Suggest how water resources and human well-being might be affected by the data in Figure 2. **(10)**

**b** Using named examples, assess the role of different players and decision-makers in trying to secure a sustainable 'water future'. **(15)**

**(Total 25 marks)**

***Note:***

*Costs are via piped household connections in the USA and UK.*
*Costs are via informal water sellers in Columbia, Philippines, and Ghana.*
*$ PPP (Purchasing Power Parity) is adjusted to reflect the cost of living in each country.*

***Sample question for Unit 3 Section B***

**6 a** Explain the factors that have led to Latin America's rapid adoption of GM farming technology. **(10)**

**b** Assess the human and environmental impacts of GM farming in Latin America. **(18)**

**c** To what extent does GM technology provide a technological fix that is economically sustainable? **(12)**

**(Total 40 marks)**

***Sample question for Unit 4***

**3** To what extent do food security issues vary spatially and temporally?

**(Total 70 marks)**

Clearly the exam questions at A2 demand rather different skills from those at AS, where the longest answers were worth 15 marks.

## Helping students to step up to A2

The biggest step up between AS and A2 is the requirement for **extended writing**. Whereas some questions at AS carry a few marks and require short answers, at A2 the questions range between 10 marks (in part (a) of Section A in the exam for Unit 3) and a full-blown 70-mark essay in Unit 4. It is important in your teaching and in setting work for students that they are given opportunities to develop extended writing skills to earn the full marks.

## Developing extended writing

Most students can write at length. The differentiator is whether they:

- 1 understand the question;
- 2 plan an answer and stay focused;
- 3 keep track of time;
- 4 learn and use case-study evidence to support an argument;
- 5 know how to organise an essay.

***I Understanding the question***

Encourage students to become thoroughly familiar with **command words**, i.e. what the examiners demand of them. The table below shows the most commonly used command words at A2. Command words at A2 tend to be different from those at AS – 'describe' for instance would rarely appear – so help students to refresh their knowledge using this list. Make flash cards or similar to help them.

| **Command word** | **What it means** | **Examples linked to topics in Unit 3** |
|---|---|---|
| Account for | Explain the reasons for – marks are given for explanation rather than description. | Account for the loss of biodiversity in coastal wetlands. |
| Analyse | Identify the main characteristics and rate the factors with respect to importance. | Analyse the social, environmental, and economic impacts of exploiting tar sands. |
| Assess | Examine closely and 'weigh up' a particular situation, e.g. strengths and weaknesses, for and against. | Using examples, assess the view that the relationship between the developed and the developing world is a neo-colonial one. |
| Comment on | This requires the assessment of a statement, putting both sides of the argument. | Comment on the view that the USA should end its dependence on oil. |
| Compare | Identify similarities *and* differences between two or more things. | Compare the strengths and weaknesses of communism and capitalism as theories. |
| Contrast | Identify the differences between two or more things. | Contrast the threats to biodiversity in two biodiversity hotspots. |
| Discuss | Similar to 'assess'. | 'Aid donors should retain greater power than the recipients.' Discuss. |
| Evaluate | The same as 'assess'. | Evaluate this statement. 'The battle between free and fair traders misses the point – it is feeding people at a price they can afford that matters.' |
| Examine | Students need to describe *and* explain. | Examine the attempts to manage the threats to the Daintree rainforest. |
| Explain | Give reasons why something happens. | Explain how the desire to develop ecosystems threatens them. |
| How far | Students need to put both sides of an argument. | How far are the conflicts in the Daintree a case of economic versus environmental interests? |
| Illustrate | Use specific examples to support a statement. | Illustrate the ways in which superpowers use resources as economic weapons. |
| Justify | Give evidence to support statements. | Rank the factors affecting access to energy and justify your rankings. |
| Outline | Students need to describe *and* explain, but give more description than explanation. | Outline the ways in which the environment influences levels of food security at regional, national, and international levels. |
| To what extent | The same as 'how far'. | To what extent do you agree with the view of the World Bank that 'many countries are poor because they do not use their water potential'? |

The purpose is to help students to read and interpret exam questions carefully, and answer the one that's set, rather than the one they hope for!

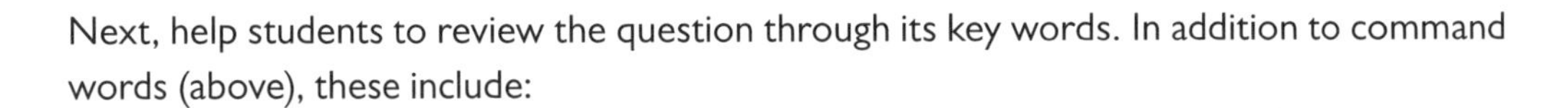

Next, help students to review the question through its key words. In addition to command words (above), these include:

**Theme or topic** – This is what the question is about. It is the means by which examiners narrow down and restrict the theme of the question so that it can be covered in 40 minutes.
**Focus** – This shows how the theme has been narrowed down, e.g. *environmental* impacts of development in China.
**Examples** – Which specific examples are required? Examples are always used at A2 – not in the kind of detail from single-case studies at AS or GCSE, but as general evidence to help support an argument.

Below is an example of a question that has been interpreted using key words on page 123.

**Command words**
- 'Explain' – Give reasons for the pattern.
- 'Suggest possible impacts' – Give more than one possible impact.
- 'Evaluate' – Examine and weigh up the relative importance of threats to the ecosystem.

**Theme or topic**
This question is for Biodiversity under threat, looking specifically at the factors and processes that threaten biodiversity.

Study Figure 3.
**a** Explain the pattern of alien species invasions, and suggest the possible impacts of alien species on ecosystems. **(10)**
**b** Evaluate the relative importance of global and local threats to one named global ecosystem. **(15)**
**(Total 25 marks)**

**Focus**
- Part (a) asks for explanations of the pattern of alien species invasions and the possible impacts this can have on ecosystems generally.
- Part (b) asks for a named example of an ecosystem and to evaluate the relative importance of threats to it.

**Examples**
Students should choose one named global ecosystem only.

***2 Planning an answer***
All the research into students who plan shows that they get more marks than those who don't. This is because:
- they stay focused – having a plan stops them from going off-track;
- they don't suffer from 'memory blanks' in which they forget what to say.

So, no matter how much consumer resistance you encounter, teach your students to plan!

Plans need not be complex or lengthy. Remember to stress to students that plans are also marked, so should never be crossed out! The recommended allocation of time for planning is 5–10% of the exam time per answer. A 35-minute answer will therefore require no more than 2–3 minutes of planning. Plans need not be elaborate either – pencil notes in the margin, used as an *aide-memoire* to keep an answer on track or to help remember detail or data, can be extremely useful.

To encourage students to plan, work on planning as part of exam preparation by giving students 2–3 minutes to plan a question, and then discuss the kinds of plans that work best for different people.

***3 Keeping to time***

So many students lose control of time in exams. Longer essays apart, most marks are earned in the first half of an answer, and prolonging it beyond its time slot will progressively earn fewer and fewer marks. It is better to draw a slightly incomplete question to a close and start a new one than to prolong. Generally, the following rules are helpful for students:

- work out how long an answer should take, and how long each sub-section should take;
- work on the basis of 5–10% planning time, 80–85% writing time, and 10% checking time.

To encourage students to keep to time, give them the exact time that they will be allowed for the exam in practice essays or questions. It is just as helpful to do this with shorter 10-mark answers (in, say, 12–13 minutes) as it is for longer essays for Unit 4. Give plenty of timed practice – and not just in the last week or two before an exam.

***4 Learning and using case-study examples***

Case studies at A2 are rather different from those used at AS and below. Occasionally at AS, one example is enough (e.g. the study of one coastal stretch under threat from erosion). At A2 a **range** of examples is required – one example of a development project in an answer about 'Bridging the Development Gap', for instance, will not normally be enough. So the Student Book includes, for example, details of progress in both Uganda and Bangladesh in trying to assess the extent to which the Millennium Development Goals are being achieved.

Sometimes, extended regions are used because they illustrate a particular requirement of the specification, such as the Daintree in Chapter 3 'Biodiversity under threat'. A question could be set such as:

> **b** Evaluate the relative importance of global and local threats to one named global ecosystem.

In this question, one case area might be sufficient, because the detail lies in the **range of threats**. If a range of threats exists in one place or region, then it may be sufficient to refer to that, especially if the question is worth only 10 or 12 marks. However, a strong answer might refer to one set of threats in one location (e.g. the Daintree), to which are added those from another (e.g. mangroves). In this particular question, students should ideally develop an answer, progressing from lesser threats to greater threats, or vice-versa.

Edexcel examiners' reports give teachers (and students!) plenty of guidance and specimen examples of good practice, so try to download and read these for each exam cycle, and attend examiners' feedback sessions as often as you can.

***A portfolio of case studies***

Students need to develop a portfolio of case studies, particularly in Unit 4. It is important for them to realise which examples they have studied fit which part of the specification. A grid like this can help them to keep track of where to fit examples and how to use them. It also helps students to make links between topics.

Topic name:

| Subject content | Relevant theory | Case studies |
|---|---|---|
| | | |
| | | |

***Remembering case studies***

Although students frequently retain good conceptual grasp of the specification, good detailed revision is sometimes hard to achieve. Revision is – or can be – boring because hours spent gazing at files prove fruitless. Encourage students to take an active approach to revision and learning detailed examples.

For example, two case studies in the Student Book would prove useful in answering the question above from 'Biodiversity under threat' – 'The Daintree rainforest' (pages 106–7) and 'Mangrove ecosystems' (pages 120–1). You could run the following exercise for preparing a rough answer.

- Ask students to draw a spider diagram with each threat, and details about it.
- Then get them to extend the diagram with ways in which threats affect the ecosystem (with examples), and identify the threats as global, national, or local.
- Finally, get them to annotate the diagram round the outside, using a large or small '+' to show how big the threat is, i.e. to help assess each threat.

You could photocopy the spider diagram below as an example, and ask students to complete it.

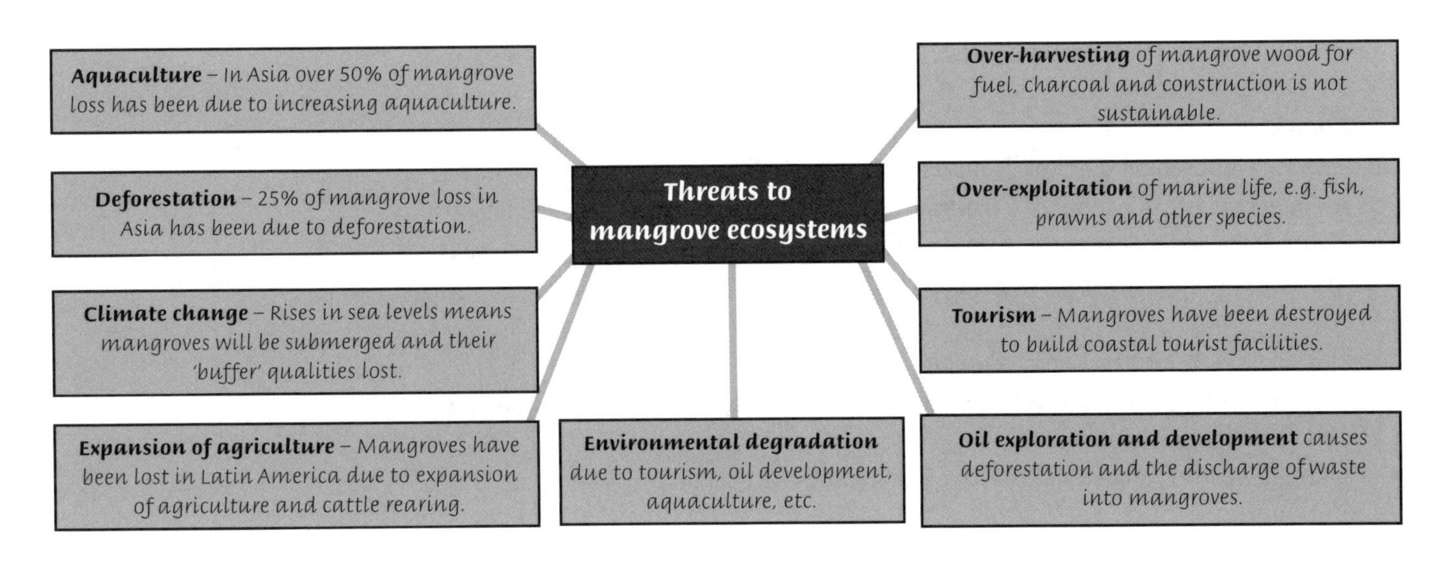


***5 Organising an essay***

Beware! Very good students can underachieve on essays. Most commonly, they show excellent knowledge and understanding, in real depth, but run out of time, or fail to answer the question. Strangely, a weaker student who actually answers the question set can do well.

Here are some handy tips to enhance students' extended writing.

***Introduce the essay***

Students should define the key terms, set out the context to which the question is referring (e.g. how far biodiversity is under threat), and outline their argument.

***Set out the main body of the essay***

Students should argue or show knowledge of both sides, not just write 'whatever comes next', and organise examples so that they:

- classify those on one side of an argument from those on the other, or into social, economic, and environmental points;
- develop, from those that are major threats to those that are minor;
- progress towards an answer, e.g. from factors that strongly support the argument to those that support it less so.

In each case, they should keep coming back to the title – 'This example shows how big a threat…', etc.

***Write a conclusion***

Students should give an answer to the question in a **conclusion**. They should answer the question fully, showing where the balance of an argument lies, or evaluating which are the major threats, for example. The trick is not to give it all away in the introduction!

## Becoming familiar with mark schemes

Part of the necessary preparation for students facing examinations lies in getting to know how the exams will be marked. The following three points are essential guidance:

- At this level, all marking is done using levels, and never points. It is therefore essential that you and your students are clear how levels are used.
- Always remember that the full range of marks is used by examiners, and that many students do get full marks.
- Remember, too, that the mark that students get on the examination – what is known as the **raw mark** – is not the mark they will see on results day. They will see a UMS mark – a Uniform Mark Scheme – which is based on the raw mark they have obtained.

***Level marking***

Level marking is based on the overall qualities of an answer, not on the number of points that are made. At AS, questions of 5 marks or fewer are point-marked; all questions with 6 marks or more are level marked. At A2, questions range from 10 marks in some parts of Unit 3 to a full-blown 70-mark single essay in Unit 4.

In Unit 3, Section A questions are in two parts:

- Part (a) is worth 10 marks. Questions with this mark allocation are marked using three levels in the mark scheme.
- Part (b) is worth 15 marks and has four levels in the mark scheme.

So, a question for Unit 3 such as:

*Study Figure 3.*

*(a) Explain the pattern of alien species invasions, and suggest the possible impacts of alien species on ecosystems.* ***(10)***

would be marked using the following mark scheme. The mark scheme is normally free text, but in fact can be broken into bullets, as shown:

| Level | Mark | Descriptor |
|---|---|---|
| Level 1 | 1–4 | • Structure is poor or absent.<br>• One or two basic ideas explaining the pattern; lacks understanding of ecosystem processes. Unlikely to describe the map.<br>• Explanations are over-simplified and lack clarity.<br>• Geographical terminology is rarely used with accuracy.<br>• Frequent grammar, punctuation and spelling errors. |
| Level 2 | 5–7 | • Structure is satisfactory.<br>• Does explain the pattern with some clarity. Some understanding of impact on ecosystems, but is incomplete; makes reference to map.<br>• Explanations are clear, but there are areas of less clarity.<br>• Lacks full range of geographical terminology.<br>• Some grammar, punctuation and spelling errors. |
| Level 3 | 8–10 | • Well-structured.<br>• Sound explanation of pattern, understanding of several processes; likely to illustrate impact on ecosystems.<br>• Descriptive language is precise.<br>• Explanations always clear.<br>• Thorough use of geographical terminology.<br>• Grammar, punctuation and spelling errors are rare. |

Note the consistency of criteria – four in total. Students need to pay attention to:
- **Structure** of the answer – that is, thinking about what to write, and placing it in order – not just 'whatever comes next';
- How well they either describe in **general** terms (low level) or offer **specific** points based upon actual evidence in the diagram (high level);
- Clarity of explanation and the degree of **understanding**;
- The use of **geographical terminology** and **quality** of grammar, punctuation and spelling.

This is a consistent approach, so that the mark schemes for the 70-mark essay questions in Unit 4 can be similarly broken down. In this unit, students answer one question only, and have to write a long essay. The 70 marks are awarded for the following five criteria:
- Introducing, defining and focusing on the question (10 marks)
- Researching and methodology (15 marks)
- Analysis, application and understanding (20 marks)
- Conclusions and evaluation (15 marks)
- Quality of written communication and sourcing (10 marks).

Each of the above is divided into four levels – even the 10-mark criteria – so to gain the most marks, students' answers need to meet the criteria for level 4.

## Being synoptic!

**Unit 3 Section B** contains the synoptic investigation of pre-released resources, and students will need to show synoptic links with other parts of the specification in order to gain maximum marks. This means that students need to:

- Refer to specific ideas or concepts that they have learned and which are implicit in the resources and exam questions.
- Make references – which need not be lengthy – where they have studied similar or related exemplars.

For example, in this part of a question from Section B:
*(b) Assess the human and environmental impacts of GM farming in Latin America.* ***(18)***

the synoptic links given in the mark scheme are about:
Unit 1.2 Migration of people to urban areas.
Unit 1.3 Questions over the costs and benefits of globalised trade.
Unit 2.3 Rural inequality – social polarisation and marginalisation of some farmers.
Unit 3.3 Biodiversity and the threat of economic development.
Unit 3.6 The neutrality of technological innovation in terms of impacts.

## Maximising marks in the exam

Gaining marks is just a matter of technique. Lower marks are often awarded not because students do not understand the material, but because they do not use or organise it fully. The following are helpful hints which depend upon students selecting the right question, case studies, the type of question they answer, and quality of communication.

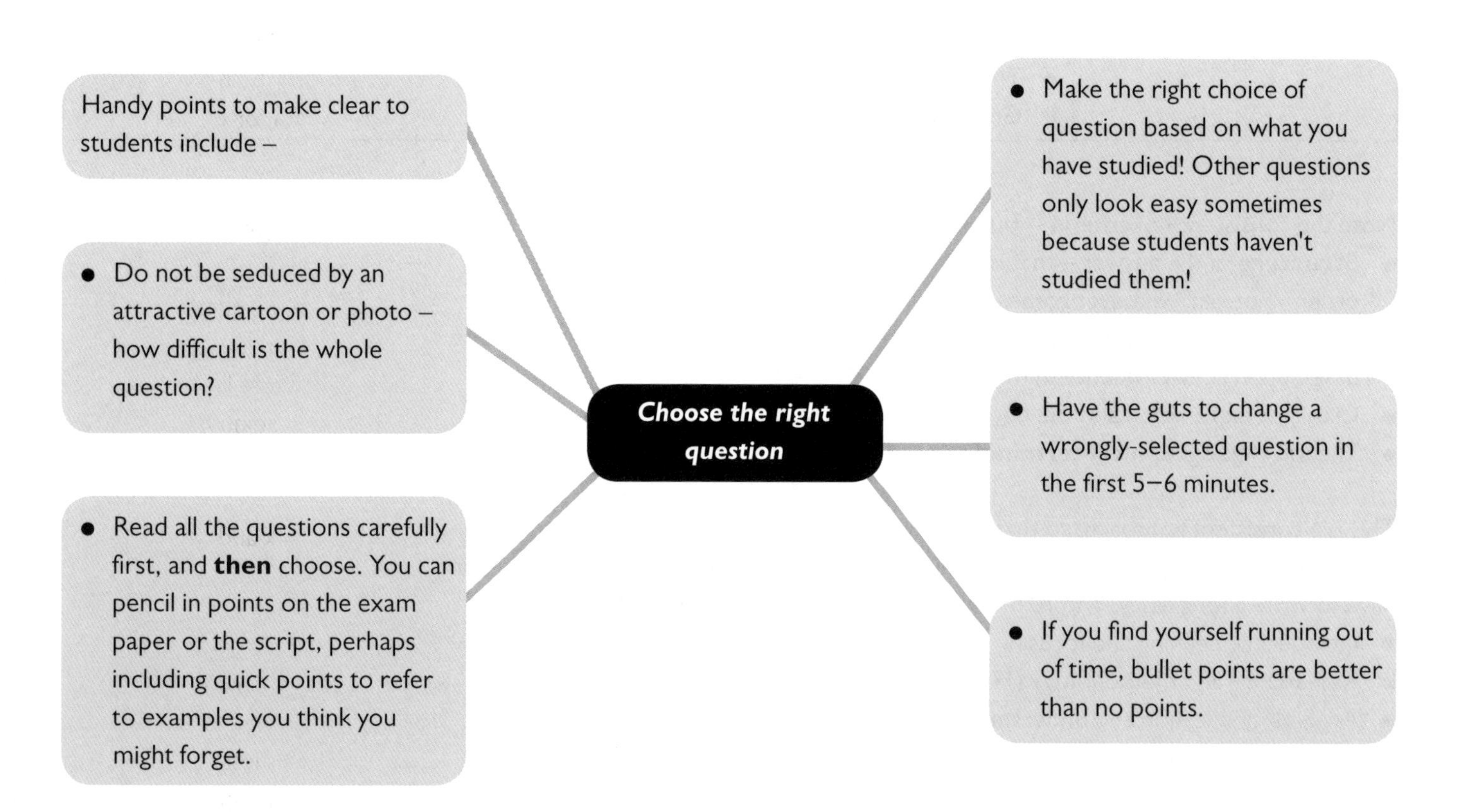

Choose the right question
Handy points to make clear to students include –
Do not be seduced by an attractive cartoon or photo – how difficult is the whole question?
Read all the questions carefully first, and **then** choose. You can pencil in points on the exam paper or the script, perhaps including quick points to refer to examples you think you might forget.
Make the right choice of question based on what you have studied! Other questions only look easy sometimes because students haven't studied them!
Have the guts to change a wrongly-selected question in the first 5–6 minutes.
If you find yourself running out of time, bullet points are better than no points.

**Choose the right case studies**

Handy points to make clear to students include –

- Choose appropriate case studies – don't include something just because you have revised it!
- Make sure you refer to more than one example. The GCSE approach is to refer to one example; AS requires perhaps two or maybe three at most; A2 may require several, especially for Unit 4. A2 requires breadth and understanding of ideas as much as detailed knowledge.

**Answer the question**

Handy points to make clear to students include –

- Read the whole question and underline the command words.
- Structure answers into sequenced paragraphs – hence the need to **plan**.
- Include keywords from the question in every paragraph.
- Use all resources provided – refer to them and the points within them. Look for and quote evidence on photos, in tables of data, or on maps.
- Sketch out a quick plan – about 1 minute maximum for every 10 marks in the question. Include examples you will use, and make a note linking each example to the question.
- Keep on track and don't go off the point.
- Use annotated maps and diagrams – a well-labelled diagram takes less time than writing a paragraph of text.

**Quality of written communication**

Handy points to make clear to students include –

- Make sure handwriting is legible. Exam boards do employ specialist staff to help out in difficult cases, and most examiners can read most handwriting! But it does not help to get the flow of the answer across if it is difficult to read.
- Use correct spelling, punctuation and grammar. Only use abbreviations where these are commonly understood, e.g. UK, USA, UN. Repeated references to, for instance, the IMF are fine provided that the International Monetary Fund (IMF) is abbreviated the first time it is referred to.
- Use a suitable style of writing; students should assume they're writing for an intelligent adult.
- Use well-planned and sequenced paragraphs.
- Use precise geographical terminology – 'migration' instead of 'people moving', for instance.
- Avoid sweeping generalisations and see the complexity in all questions. At this level, the person who questions the question is performing at the top level.

# Glossary

**A**

**altruism** – benefiting other persons or groups of people. Caring about the needs and happiness of other people more than your own

**anaerobic** – any organism or process which can or must exist without free oxygen from the air

**antiretroviral drugs** – drugs which inhibit the reproduction of retroviruses (viruses composed of RNA rather than DNA). The best known of this group is HIV, human immunodeficiency virus, which causes AIDS

**aquifer** – a rock, such as chalk, which will hold water and let it through

**areal** – relating to a roughly bounded part of the space on a surface; a region

**B**

**bilateral aid** – foreign aid (in the shape of money, expertise, education or technology) from a single donor to a recipient country

**biofuel** – solid, liquid or gas fuel derived from relatively recently dead biological material. This includes ethanol, diesel or other liquid fuels made from processing plant material or waste oil. But it also includes wood, or liquids made from wood and biogas (methane) from animals' excrement

**biomanufacturing** – concerned with the manufacture of biopharmaceuticals, which are medical drugs produced using biotechnology – the modification of biological organisms, largely through genetic engineering

**biome** – a naturally occurring community, characterised by distinctive life forms which are adapted to the broad climatic type

**British Overseas Territory** – fourteen territories that are under the sovereignty of the United Kingdom, but which do not form part of the United Kingdom itself. They are mostly small islands, mainly in the Caribbean and the South Atlantic

**C**

**carrying capacity** – the maximum population that a particular environment can sustain without incurring environmental damage

**Cold War** – this term is used to describe the relationship between America and the Soviet Union from 1945 to the late 1980s. Neither side ever fought the other – both countries had nuclear weapons and the consequences would have been appalling – but they did 'fight' for their beliefs using 'client states' who fought on their behalf, e.g. South Vietnam was anti-communist and was supported by America, while North Vietnam was pro-communist and fought the south (and the Americans) using weapons from communist Russia and communist China

**command government** – where macro-economic policy and entrepreneurial activity is controlled by the State. The Government decides what to produce, how to produce it and who to produce it for – but with some freedom for individual decisions

**commodity** – a product or a raw material that can be bought and sold, especially between countries

**commodity trading exchanges** – these are where various commodities (i.e. raw or primary products) are traded. Most commodity markets across the world trade in agricultural products and other raw materials (like wheat, barley, sugar, maize, cotton, cocoa, coffee, milk products, pork bellies, oil, metals, etc.) and contracts based on them

**conservation** – the protection of natural or man-made resources for later use

**coral bleaching** – the loss of colour of corals. Under stress, such as a change in water temperature or a bacterial infection, coral will expel the symbiotic unicellular algae which give it its colour

**D**

**desalination** – the conversion of salt water into fresh water by the partial or complete extraction of dissolved solids

**devaluation** – the reduction in the value of the currency of one country when it is exchanged for the money of another country

**digital blackout** – where people are without some or all of the following: e-mail, the Internet, television and telephone connections. This may be because of a malfunction or because of a switch to new technology

**disparity** – a difference, especially one connected with unfair treatment

**DNA** – the chemical in the cells of animals and plants that carries genetic information. It is a type of nucleic acid

**E**

**ecosystem diversity** – the concept that biodiversity (i.e. the varied range of flora and fauna) is essential for the functioning and/or sustainability of an ecosystem

**El Niño** – a southerly warm ocean current, which develops off the coast of Ecuador about fourteen times a century. It is associated with major variations in tropical climates

**electronic colonialism** – this is a theory that electronic mass media (film, TV, the Internet and commercials) are having an impact on the minds of audiences around the world, making them think along the same lines and receptive to the same influences. This 'electronic empire' is not based on military power or land acquisition, but on controlling the mind – and the English language is the means by which it achieves its ends

**energy poverty** – when a country or region has insufficient access to reliable sources of power

**energy surplus** – when a country or region has more than enough sources of power for its needs and is able to export its surplus power to other countries

**environmental determinism** – the view that human

activities are governed by the physical environment. People and peoples are what they are because they have been shaped by their physical surroundings – climate, vegetation, etc.

**erratic** – a large boulder which has been transported by a glacier so that it has come to rest on country rock of a different composition and structure

**Euro-centric** – seeing things from a European point of view

**evangelism** – persuading people to become Christians by building churches, preaching and setting up schools

## F

**formal economy** – the economy that is regulated by the State. It is economic activity that is taxed and monitored by the Government; and is included in the Government's Gross National Product (GNP)

**futures market** – this is where futures are traded. A future is a financial contract obligating the buyer to purchase an asset (or the seller to sell an asset), such as a physical commodity, at a predetermined future date and price

## G

**genetic diversity** – the theory that genetic diversity and biodiversity are dependent on each other – that diversity within a species is necessary to maintain diversity among species, and vice versa.

**genetic modification** – the manipulation of DNA by splitting the DNA molecule and then rejoining it to form a hybrid molecule. This technique bypasses biological restraints to genetic exchange and is used in agriculture to produce 'transgenic' plants, which have greater resistance to pests, herbicides or harsh environmental conditions

**Gini co-efficient** – a measure of the difference between a given distribution of some variable, like population or income, and a perfectly even distribution

**glasnost** – a Russian word meaning 'openness'. Policies based on this idea were introduced into the USSR by Mikhail Gorbachev between 1985 and 1990, as part of his wider policy of perestroika. By cultivating a spirit of intellectual and cultural openness, which encouraged public debate and participation, Gorbachev hoped to increase the Soviet people's support for and participation in perestroika.

**Green Revolution** – the transformation of agriculture that began in 1945 when agricultural research led to the development of more productive varieties of wheat in order to feed the rapidly growing populations of third world countries. The consensus among some agronomists is that the Green Revolution has allowed food production to keep pace with worldwide population growth

**groundwater** – all water found under the surface of the ground which is not chemically combined with any minerals present, but not including underground streams

## H

**hegemon** – the controlling country or organization within a particular grouping or confederacy

**high-pressure system** – a region of high atmospheric pressure, otherwise known as an 'anticyclone'. In Britain the term is generally applied to pressures of over 1000 mb

**high-technology** – the most modern methods and machines, especially electronic ones

## I

**ideology** – a set of ideas that an economic or political system is based on, or a set of beliefs that influences the way people behave

**infiltration** – the process of water entering rocks or soil

**informal economy** – all economic activities that fall outside the formal economy regulated by the State. It is economic activity that is neither taxed nor monitored by the Government; and is not included in the Government's Gross National Product (GNP)

**information and communications technology (ICT)** – this is a blanket term to cover all technologies involved in the manipulation and communication of information

**infrastructure** – the basic systems and services that are necessary for a country or an organization

**intermediate technology** – tools and technology for developing countries that are significantly more effective and expensive than traditional methods, but still much cheaper than developed world technology. Such items can be easily bought and used by poor people, and can lead to greater productivity without creating social dislocation. Much intermediate technology can also be built and serviced using locally available materials and knowledge

**inter-tidal area** – otherwise known as the 'inter-tidal zone' or the 'foreshore', is the area that is exposed to the air at low tide and submerged at high tide, for example, the area between tide marks

**irrigation** – the supply of water to the land by means of channels, streams and sprinklers in order to permit the growth of crops

## L

**La Niña** – an extensive cooling of the central and eastern Pacific. Globally La Niña means that parts of the world that normally experience dry weather will be drier and those with wet weather will be wetter. Typically La Niña will last for up to 12 months and will be a less-damaging event than the stronger El Niño

**lenticels** – spongy areas in the corky surface of plant parts, such as twigs and stems, that allow gas exchange between the atmosphere and the internal tissues of the plant. In some plant species, lenticels are raised round dots; they may also be seen as vertical or horizontal slits

**low-carbon standard** – an initiative first introduced in California in 2007, which is aimed at reducing the carbon intensity of transportation fuel by 10% by 2020. Fuel providers can choose how to achieve this target by various means, including blending low-carbon ethanol into petrol, buying credits from utilities supplying electricity to electric cars, and diversifying into low-carbon hydrogen as a fuel for motor vehicles

**M**

**moraine** – any landform directly deposited by a glacier or ice sheet
**multilateral aid** – foreign aid (in the shape of money, expertise, education or technology) from a group of countries or donors to a single country
**multiplier effect** – an effect in economics in which an increase in spending produces an increase in national income and consumption greater than the initial amount spent

**N**

**nanoparticle** – a small object, used in nanotechnology, that behaves as a whole unit in terms of its transport and properties. It is sized between 1 and 100 nanometres
**nationalist** – a person (or a political party) who feels that their country should be independent and has a great love and pride for their country. But it can also mean people who think their country is better than any other
**neo-liberalism** – the doctrine that market exchange is an ethic in itself, capable of acting as a guide for all human action. Under neo-liberalism, State interventions in the economy are minimized, while the obligations of the State to provide for the welfare of its citizens are diminished
**net primary productivity** – primary production is the production of chemical energy in organic compounds by living organisms. Net primary production is the rate at which all the plants in an ecosystem produce net useful chemical energy. Some net primary production will go towards growth and reproduction of primary producers, while some will be consumed by herbivores
**Non-Governmental Organisation (NGO)** – a legal organisation created by private organisations or people with no participation or representation by any government. Where NGOs are funded by governments, the NGO maintains its non-governmental status by excluding Government representatives from membership in the organisation

**P**

**pandemic** – a disease that spreads over a whole country or over the whole world
**pasteurisation** – where a liquid, especially milk, is heated to a particular temperature and then cooled in order to kill harmful bacteria
**percolation** – the filtering of water downwards through soil and through the bedding planes, joints and pores of a permeable rock
**perestroika** – the policy of economic and governmental reform instituted by Mikhail Gorbachev in the Soviet Union during the mid-1980s. It is a Russian word meaning 'restructuring'
**periglacial environments** – arid areas with a tundra climate where temperatures are below 0 °C for at least six months, and summers are warm enough to allow surface melting to a depth of approximately one metre
**permafrost** – areas of rock and soil where temperatures have been below freezing point for at least two years. Permafrost does not have to contain ice – a sub-zero temperature is the sole qualification
**pervasive** – existing in all parts, or spreading gradually to affect all parts. Technology is spreading everywhere pervasively
**photosynthesis** – the chemical process by which green plants make organic compounds from atmospheric carbon dioxide and water, in the presence of sunlight
**phreatic eruption** – a volcanic eruption where meteoric water (water precipitated from the atmosphere) is mixed with lava
**pingo** – a large ice mound formed under periglacial conditions from an unfrozen pocket confined by approaching permafrost, possibly on a former lake. When the inner ice lens melts, the pingo collapses leaving a depression surrounded by ramparts
**pneumatophores** – specialized aerial roots which enable plants to breathe air in habitats that have waterlogged soil. The roots may grow down from the stem, or up from typical roots
**poverty line** – the official level of income that is necessary to be able to buy the basic things you need such as food and clothes and to pay for somewhere to live
**precipitation** – the deposition of moisture from the atmosphere onto the Earth's surface. This may be in the form of rain, hail, frost, sleet or snow
**prevailing** – most frequent, or most common
**privatisation** – the sale of a business or industry so that it is no longer owned by the Government
**prop roots** – modified roots that grow from the lower part of a stem or trunk down to the ground, providing a plant with extra support
**pyroclastic flows** – the result of the bursting of gas bubbles within the magma during a volcanic eruption. Lava is fragmented and a dense cloud of fragments is thrown out (a mixture of hot gases, volcanic fragments, ash, and pumice). Pyroclastic flows are also known as 'nuees ardentes'

**R**

**rain shadow** – an area of relatively low rainfall to the lee (sheltered from the wind) side of uplands. The incoming air has been forced to rise over the highland, causing precipitation on the windward side
**reciprocity** – a situation in which two countries or people provide the same help or advantages to each other
**relict features** – a geomorphological feature which existed under past climatic regimes but still exists as an anomaly in the changed, present-day conditions
**relief rainfall** – this forms when moisture-laden air masses

are forced to rise over high ground. The air is cooled, the water vapour condenses, and precipitation occurs
**riparian** – relating to a river bank. Owners of land crossed or bounded by a river have 'riparian rights' to use the river

## S

**security premium** – the extra cost built into the price of oil to allow for any disruption in supply
**segregation** – the act or policy of separating people of different races, religions or sexes and treating them differently
**seismic** – of an earthquake
**spatial imbalance** – the uneven distribution or location across a landscape or surface of, for example, population
**species diversity** – a measure of the diversity within an ecological community that incorporates both species richness (the number of species in a community) and the evenness of species' abundance
**strategic** – something that is done as part of a plan that is meant to achieve a particular purpose or to gain an advantage
**streamflow** – the flow of water in streams, rivers, and other channels. It is a major element of the water cycle and the main mechanism by which water moves from the land to the oceans
**subsidiary companies** – companies that are controlled by a bigger and more powerful company
**surface runoff** – the movement over ground of rainwater. It occurs when the rainfall is very heavy and when the rocks and soil can absorb no more
**sustainable use** – a strategy of reducing the environmental impacts associated with resource use and to do so in a growing economy. This would require changes in consumption patterns, better education, new technologies and higher prices for exploiting ecosystems

## T

**tariff** – a tax that is paid on goods coming into or going out of a country
**technology poor** – places and people who lack access to a regular and reliable source of electricity are technology poor. Where there is electricity, there may not be access to the Internet. The gap between digital 'haves' and 'have-nots' is sometimes referred to as the 'digital divide'
**technology rich** – places and people who have access to reliable electricity and to a good communications infrastructure are technology rich
**tied aid** – where foreign aid benefits the donor in the shape of interest payments, access to new markets or by political allegiance. Tied aid sometimes has to be spent in the country providing the aid (the donor country), or in a group of selected countries

## U

**urbanisation** – the migration of rural populations into towns and cities. It indicates a change of employment structure from agriculture and cottage industries to mass production and service industries

## V

**value-added** – the additional value of a good over the cost of commodities used to produce it from the previous stage of production. It refers to the contribution of the factors of production, i.e. land, labour, and capital goods, to raising the value of a product

## W

**water rights** – the legal right of a user to use water from a water source, e.g. a river, stream, pond or source of groundwater
**water wars** – international conflict as a result of pressure on water supplies. This is an increasing possibility as a growing and increasingly urbanised global population will increase demand for food and water, at the same time as climate change and other trends put greater pressure on their supply
**world water gap** – the difference between those people, nearly two-thirds of the world's population, who live in water poverty (with either a physical or economic scarcity of water, or both) and those who have ready and reliable access to water for drinking and sanitation

## Z

**zonation** – The distribution of organisms in biogeographic zones